T3-BHV-003

Editor-in-Chief: **Lord Walton**

The Use of Isradipine and Other Calcium Antagonists in Cardiovascular Diseases

Proceedings of a symposium sponsored by
Sandoz BV and held in Amsterdam, 20 January 1989

Edited by

P. A. van Zwieten

ROYAL SOCIETY OF MEDICINE SERVICES
LONDON · NEW YORK
1989

Royal Society of Medicine Services Limited
1 Wimpole Street   London W1M 8AE
7 East 60th Street   New York NY 10022

©1989 Royal Society of Medicine Services Limited

All rights reserved. No part of this book may be reproduced in any form by photostat, microfilm, or any other means, without written permission from the publishers.

This publication is copyright under the Berne Convention and the International Copyright Convention. All rights reserved. Apart from any fair dealing under the UK Copyright Act 1956, Part 1, Section 7, no part of this publication may be reproduced, stored in a retrieval system or transmitted in any form or by any means without the prior permission of the Honorary Editors, Royal Society of Medicine.

These proceedings are published by Royal Society of Medicine Services Ltd with financial support from the sponsor. The contributors are responsible for the scientific content and for the views expressed, which are not necessarily those of the sponsor, of the editor of the series or of the volume, of the Royal Society of Medicine or of Royal Society of Medicine Services Ltd. Distribution has been in accordance with the wishes of the sponsor but a copy is available to any Fellow of the Society at a privileged price.

British Library Cataloguing in Publication Data
The use of isradipine and other calcium antagonists in
  cardiovascular diseases
  1. Medicine. Drug therapy. Calcium antagonists
  I. Zwieten, P. A. Van   II. Series
  615'.7

ISBN 1-85315-108-4

Phototypeset by Dobbie Typesetting Limited, Plymouth, Devon
Printed in Great Britain at the Alden Press, Oxford

Contributors

Editor

P A. van Zwieten   *Departments of Pharmacotherapy and Cardiology, Academic Medical Centre, University of Amsterdam, The Netherlands*

Contributors

J. Beekman   *Department of Internal Medicine, Medisch Spectrum Twente, Ariënsplein 1, 7511 JX Enschede, The Netherlands*

J. H. B. de Bruijn   *Department of Internal Medicine, Medisch Spectrum Twente, Ariënsplein 1, 7511 JX Enschede, The Netherlands*

W. Diemont   *Department of Internal Medicine, Medisch Spectrum Twente, Ariënsplein 1, 7511 JX Enschede, The Netherlands*

W. H. van Gilst   *Department of Cardiology, Thoraxcenter, Groningen, The Netherlands*

A. Hof   *Preclinical Research Department, Sandoz Ltd, CH-4002 Basel, Switzerland*

R. P. Hof   *Preclinical Research Department, Sandoz Ltd, CH-4002 Basel, Switzerland*

G. Leonetti   *Istituto di Clinica Medica Generale e Terapia Medica, Università di Milano and Centro di Fisiologia Clinica e Ipertensione, Ospedale Maggiore, Milano, Italy*

K. I. Lie   *Department of Cardiology, Thoraxcenter, Groningen, The Netherlands*

A. J. Man in 't Veld   *Department of Internal Medicine 1, University Hospital Dijkzigt, Erasmus University Rotterdam, The Netherlands*

M. W. J. Messing   *Department of Pharmacology, University of Limburg, Maastricht, The Netherlands*

M. Rudin   *Preclinical Research Department, Sandoz Ltd, CH-4002 Basel, Switzerland*

U. T. Rüegg   *Preclinical Research Department, Sandoz Ltd, CH-4002 Basel, Switzerland*

A. Sauter   *Preclinical Research Department, Sandoz Ltd, CH-4002 Basel, Switzerland*

A. M. J. Siegers   *Department of Internal Medicine, Medisch Spectrum Twente, Ariënsplein 1, 7511 JX Enschede, The Netherlands*

H. A. J. Struyker Boudier
Department of Pharmacology, University of Limburg, Maastricht, The Netherlands

T. Thien   *Department of Medicine, Division of General Internal Medicine, St Radboud University Hospital, Geert Grooteplein Zuid 8, 6500 HB Nijmegen, The Netherlands*

E. W. van den Toren
Department of Cardiology, Thoraxcenter, Groningen, The Netherlands

D. B. Weinstein   *Department of Lipid and Lipoprotein Metabolism, Sandoz Research Institute, Route 10, East Hanover, New Jersey 07936, USA*

H. Wollersheim   *Department of Medicine, Division of General Internal Medicine, St Radbound University Hospital, Geert Grooteplein Zuid 8, 6500 HB Nijmegen, The Netherlands*

A. Zanchetti   *Istituto di Clinica Medica Generale e Terapia Medica, Università di Milano and Centro di Fisiologia Clinica e Ipertensione, Ospedale Maggiore, Milano, Italy*

Contents

Contents

Preface

P. A. van Zwieten

Departments of Pharmacotherapy and Cardiology,
Academic Medical Centre, University of Amsterdam, The Netherlands

Both the fundamental and the clinical interest for calcium antagonistic drugs are continuing at a high level and our insight in this subject is advancing at a rapid pace, even after two decennia of intensive research in this field. This ongoing interest is also paralleled by the continuous design and introduction of new drugs of this category.

Isradipine (PN 200–110) is such a new calcium antagonist, which belongs to the major subgroup of the dihydropyridines, of which nifedipine, nitrendipine and nimodipine are well-known examples.

After an initial meeting on calcium antagonists in 1987 a subsequent Round Table Conference was organized in 1989, again in Amsterdam, on 'The use of isradipine (PN 200–110, Lomir®) and other calcium antagonists in cardiovascular diseases'. Several experts in the field critically discussed the use of calcium antagonists in cardiovascular diseases in general, and that of isradipine in that particular context.

Accordingly, an up to date survey of the subject was obtained, as reflected in this publication. Similarly as for the first Round Table Conference on this subject I had great pleasure in chairing this recent conference of this type and in editing the concomitant proceedings, published in this monograph. Also on behalf of the participants I would like to gratefully acknowledge the generous hospitality of the Sandoz-Nederland Company (Uden, The Netherlands), which allowed us to communicate freely and to leave the scientific responsibility entirely to the chairman and speakers.

New developments in the field of calcium antagonists

P. A. van Zwieten

Departments of Pharmacotherapy and Cardiology,
Academic Medical Centre, University of Amsterdam, The Netherlands

INTRODUCTION

Although calcium antagonists have been widely recognized as established therapy in cardiovascular disease for many years, the clinical and fundamental interest in these drugs has by no means diminished. Before going into detail concerning the new compound isradipine (PN 200-110) it seems useful briefly to emphasize a few recent developments in the field. These new developments and aspects cover rather heterogeneous fields, as for instance definitions and nomenclature, the WHO classification of calcium antagonists, differentiation of calcium channels, the introduction of intracellularly acting calcium antagonists and new therapeutic indications of existing calcium antagonists, in particular in the treatment of neurological disorders. This brief introductory chapter does not aim at reviewing the state of the art of the field of calcium antagonistic drugs, but is intended only to put forward a few of the aforementioned new aspects. With respect to the state of the art of existing calcium antagonists and their applications we recommend the monograph 'Calcium-antagonists' edited by Vanhoutte et al. (1988) (1), representing a high-level and virtually complete survey of the field.

DEFINITIONS AND NOMENCLATURE

After a variety of definitions of calcium antagonists, submitted in the more remote and recent past, we would like to adhere to the definition recently proposed by Godfraind (2).

'A calcium antagonist is a compound that alters the cellular function of calcium by inhibiting its entry and/or its release by interfering with one of its intracellular actions'.

This definition covers both the calcium channel blockers, which predominantly act at the level of the membrane and the compounds with an intracellular action. Even the calcium overload blockers (3), which act only or predominantly under

The use of isradipine and other calcium antagonists in cardiovascular diseases, edited by P. A. van Zwieten, 1989; Royal Society of Medicine Services International Congress and Symposium Series No. 157, published by Royal Society of Medicine Services Limited.

1

pathological, ischaemic conditions are covered by Godfraind's definition. Even the calmodulin antagonists are included in this definition (4).

In accordance with the recently established and generally accepted WHO classification of the calcium antagonists the only generally applicable nomenclature of these compounds remains indeed limited to *calcium antagonists*. Although there are sound arguments in defence of terms like calcium entry blockers, calcium channel blockers, calcium slow channel blockers etc. we shall, throughout this publication, use the term *calcium antagonists* and we defend (although sometimes reluctantly) its general use throughout the literature at present and in the future.

THE WHO CLASSIFICATION OF CALCIUM ANTAGONISTS

After many attempts to subclassify the particularly heterogeneous group of calcium antagonists, the recently introduced WHO classification appears to be a workable and useful base for dealing with the heterogeneity of the various compounds involved. The WHO classification may be summarized as follows (5):

    class   I: phenylalkylamines (verapamil-like)
    class II: dihydropyridines (nifedipine-like)
    class III: benzothiazepines (diltiazem-like)
    class IV: diphenylalkylamines (flunarizine and related compounds).

This subdivision largely follows the chemical structure of the various groups of compounds. The WHO classification will require revision and/or extension when new types of calcium antagonists with as yet unknown chemical structures are introduced.

DIFFERENTIATION OF CALCIUM CHANNELS

It is widely accepted that the specific calcium channels in the cell membranes of various tissues are heterogeneous, thus requiring differentiation between at least the potential operated and receptor operated channels (POC and ROC), respectively. Initially the POC were assumed to be the only target of the calcium antagonists, but more and more evidence has accumulated which supports a relevant role for the ROC as target of the calcium antagonists as well. The work of our own group (6,7) strongly suggests that predominantly alpha-2-adrenoceptors are functionally coupled to the ROC, although under certain circumstances alpha-1-adrenoceptors may play a role as well as the receptors triggering the opening of the ROC. Recently we proposed (8) that angiotensin II receptors might also be involved in triggering the ROC.

A more refined differentiation of the POC has recently been submitted (9). This differentiation is based upon the membrane potential which is required to activate the POC of cardiac cells. Accordingly, the T-(transient) channels require relatively negative membrane potentials (-20 to -70 mV) for their activation, while L-(large conduction) channels activate at more positive potentials in the range -50 to $+10$ mV. Only the L-channel is sensitive to calcium antagonists and also to calcium promoters such as compound Bay k 8644. A further division of the L-channels into various subtypes is also a subject of debate.

It should be emphasized that the differentiation into T- and L-channels has only been investigated in cardiac tissues. In neuronal tissues a third type of channel has been demonstrated, i.e. the N-channel, which is distinct from cardiac T- and L-channels.

It may be supposed that the differentiation between the various calcium channel subtypes may lead to the development of more selective calcium antagonists, specifically interacting with each particular type of calcium channel.

Intracellularly acting calcium antagonists are assumed to impair the release of calcium ions from intracellular particles and/or calcium pools. TMB_8 is one example of the several types of experimental compounds developed so far. Clinical data on these compounds are not available yet and their potential therapeutic use remains an open question.

NEW THERAPEUTIC INDICATIONS OF THE EXISTING CALCIUM ANTAGONISTS

In cardiovascular diseases calcium antagonists have become widely recognized therapeutic agents in angina pectoris, hypertension, supraventricular tachycardia and certain types of cardiomyopathy. In animal models dihydropyridines and also verapamil have been shown to inhibit atherogenesis, but this finding has still to be extended to human studies, which are notoriously difficult to perform. It would be of great value to possess drugs which may impair atherogenesis in humans, but this goal remains remote.

A new field has been opened by means of the application of calcium antagonists in certain neurological disorders. In particular nimodipine, a dihydropyridine (WHO class II) and the diphenylalkylamine flunarizine (WHO class IV) have been investigated for therapeutic potential in neurological disorders. The position of these drugs may be globally summarized as follows (10):

(a) nimodipine has been shown to improve the sequelae of acute cerebral ischaemia, although a debate continues on whether its vasodilator action may be dangerous and cause a steal phenomenon;
nimodipine has been shown to improve both survival and neurological damage following a subarachnoid haemorrhage;
nimodipine is subject to investigation as a potential therapeutic in migraine;
(b) flunarizine is an established antimigraine drug; certain forms of vertigo, especially that associated with vertebro-basilar insufficiency;
flunarizine may be effective as an add-on remedy in epilepsy, although this has not been confirmed in all studies;
i.v. flunarizine is being tested in the acute treatment of stroke.

It is believed that the favourable effect of nimodipine in stroke is mainly caused by an anti-ischaemic effect at a cellular level, whereas in subarachnoid haemorrhage relaxation of cerebrovascular spasm is the cause of its therapeutic effect.

Flunarizine is believed to act mainly via an anti-ischaemic effect, vasodilatation not being an important basis of its therapeutic potency.

However, it should be realized that no direct evidence is available that the therapeutic efficacy of calcium antagonists in certain neurological disorders is directly related to the calcium antagonistic activity of these compounds. Whatever their mechanism in neurological therapy may be, the calcium antagonists have opened a few promising therapeutic possibilities in this difficult field.

CONCLUSIONS AND PERSPECTIVES

The heterogeneity of the various calcium antagonists so far developed has been largely confirmed by the sophisticated techniques which allow the differentiation of calcium antagonistic mechanisms and calcium channels. This heterogeneity is also confirmed by the wide spectrum of established and potential therapeutic applications of these drugs. Although hardly the subject of the present monograph, the application of calcium antagonists, which advances at a rapid pace, is certainly a fascinating development. This wide spectrum of applicability, as well as the sophisticated differences in the mode of action at a cellular level, justify the hope that even more selective compounds, with more precise therapeutic targets, can be developed. The potential success of such compounds would largely compensate for the inconvenience of adjusting and extending the recently issued WHO classification of calcium antagonistic drugs.

REFERENCES

(1) Vanhoutte PM, Paoletti R, Govoni S (eds.). Calcium antagonists; pharmacology and clinical research. *Ann NY Acad Sci* 1988; **522**: 1–802.
(2) Godfraind Th, Miller R, Wibo M. Calcium antagonism and calcium entry blockade. *Pharm Rev* 1986; **38**:321–46.
(3) Van Zwieten PA. Differentiation of calcium entry blockers into calcium channel blockers and calcium overload blockers. *Eur Neurol* 1986; **25** (Suppl 1): 56–67.
(4) Hidaka H, Tanaka T. Modulation of Ca^{2+}-dependent regulatory systems by calmodulin antagonists and other agents. In: Hidaka H, Hartshorne DJ, eds. *Calmodulin antagonists and cellular physiology*. New York: Academic Press, 1985: 13–23.
(5) Vanhoutte PM, Paoletti R. The WHO classification of calcium antagonists. *Trends Pharmacol Sci* 1987; **8**:4–5.
(6) Van Zwieten PA, Timmermans PBMWM, Thoolen MJMC, Wilffert B, de Jonge A. Calcium dependency of vasoconstriction mediated by α_1- and α_2-adrenoceptors. *J Cardiovasc Pharmacol* 1985; **7** (Suppl 6): S113–20.
(7) Van Zwieten PA, Timmermans PBMWM. Alpha-adrenoceptor stimulation and calcium movements. *Blood Vessels* 1987; **24**: 271–80.
(8) Van Zwieten PA, Timmermans PBMWM, van Heiningen PNM. Receptor subtypes involved in the action of calcium antagonists. *J Hypertension* 1987; 5(Suppl 4): S21–8.
(9) Nowycky MC, Fox AP, Tsien RW. Three types of calcium channels in chick dorsal root ganglion cells. *Biophys J* 1985; **47**: 67a.
(10) Van Zwieten PA. The use of calcium entry blockers in the treatment of neurological disorders. *Progr Clin Neurosci* 1986; **2**: 33–40.

The pharmacological basis for new therapeutic uses of isradipine

R. P. Hof, U. T. Rüegg, A. Hof, A. Sauter and M. Rudin

Preclinical Research Department, Sandoz Ltd, CH-4002 Basel, Switzerland

SUMMARY

Calcium antagonists share a variety of pharmacodynamic actions which, however, may manifest themselves at widely different doses in intact animals or man. Therefore, calcium antagonists vary in their pharmacological profile and may be preferentially suited for certain therapeutic uses. Isradipine is well known under the code name of PN 200-110 (PN) for its high affinity and selectivity for the dihydropyridine binding site of the slow calcium channel. It shows pronounced target tissue selectivity in experiments *in vitro* and *in vivo*. Thus isradipine slows the sinus node at doses that do not inhibit A-V-nodal conduction and myocardial contractility. In a rabbit model imitating some aspects of heart failure isradipine was remarkably free of cardiodepressant activity, differing thus from nifedipine or diltiazem. Isradipine may be suitable as an afterload reducing agent for the treatment of heart failure (see also below).

The inhibition of the L-channel on vascular smooth muscle is the mechanism of the potent vasodilator and antivasoconstrictor effect of isradipine. Interestingly the antivasoconstrictor effect was enhanced in atherosclerotic rabbits, suggesting that isradipine may compensate to some extent the defective endothelial function in these animals. The blood pressure fall was also more pronounced in the atherosclerotic group than in the control animals. These findings show that isradipine not only has an antiatherosclerotic action, but that there are also immediate functional benefits in such animals suggesting that isradipine might be especially suited for vasodilator therapy in patients with atherosclerosis.

There is evidence that the sympathetic system and the renin–angiotensin system may be activated by and interfere with some actions of calcium antagonists, a fact which should guide the selection of concomitant therapy, especially for heart failure, but also for hypertension. From a pharmacological point of view, a combination with diuretics and nonspecific vasodilators, which have similar effects on these systems, appears less desirable than a combination with agents, which oppose these effects. Pharmacological experiments, again in rabbits, favour angiotensin-converting enzyme inhibitors or beta-blocking agents (for the treatment of hypertension) as candidates for a combination therapy with isradipine.

The use of isradipine and other calcium antagonists in cardiovascular diseases, edited by P. A. van Zwieten, 1989; Royal Society of Medicine Services International Congress and Symposium Series No. 157, published by Royal Society of Medicine Services Limited.

In a rat model of stroke, where the infarct size was followed with nuclear magnetic resonance imaging, isradipine decreased infarct size by up to 60%, if therapy was started immediately after occlusion of the cerebral artery. The protective effect was more pronounced than that of nimodipine, nitrendipine, darodipine and nicardipine. The protection was dose-related and also observed after a delayed onset of treatment, even though the degree of protection diminished with the duration of the delay.

Besides the well established use of isradipine in hypertension, animal experiments thus suggest potential new uses in stroke and heart failure. Furthermore, atherosclerosis induced hyperresponsiveness to vasoconstrictor agents responds well to treatment with isradipine.

ISRADIPINE AND THE MULTIPLE PHARMACODYNAMIC ACTIONS OF CALCIUM ANTAGONISTS

Calcium antagonists share the mechanism of action, blockade of the slow calcium channel or L-channel, yet nevertheless show differing patterns of cardiovascular activity. Thus verapamil elicits a comparatively strong negative inotropic effect, nifedipine a prominent vasodilator action and diltiazem a relatively selective negative chronotropic effect (1–3). Such differences were elaborated and illustrated well by the careful comparative studies performed in the laboratory of Taira (4). They showed that all calcium antagonists tested were coronary vasodilators, but that effects on other cardiac tissues varied. Thus verapamil and diltiazem both inhibited the sinus and atrio-ventricular node function but verapamil was more cardiodepressant than diltiazem (4). Similar differences were also observed within the class of dihydropyridine calcium antagonists (4), to which isradipine belongs, (code-name PN200-110, trade names Lomir, Dynacirc, Prescal). Such differences, demonstrated with relative ease in pharmacological experiments, might well be of therapeutic importance. So far this calcium antagonist has been characterized in clinical studies as an antihypertensive agent. This review will explore the pharmacological evidence indicating that the unique pattern of pharmacodynamic activity might make isradipine suited for uses other than the traditional indications of calcium antagonists.

The molecular basis for the tissue selectivity of calcium antagonists is still not well understood. All Ca^{++}-antagonists used at present act only on one of the calcium channels, the L-channel (5,6). This channel has several distinct binding sites (7–9). It is not yet known, however, whether the L-channels in different tissues are structurally different and whether they invariably have all the binding sites. The action of calcium antagonists is also dependent on the membrane potential. It is becoming increasingly clear that the membrane potential influences the action of individual drugs differently (5,10–12).

The multiplicity of actions together with the possibility to select compounds with a certain degree of tissue selectivity made calcium antagonists a very interesting subject for a drug design program. The primary goal in our early program was to obtain compounds that, like diltiazem or verapamil, inhibited the sinus node, but were less cardiodepressant than these agents. They should furthermore possess the potent vasodilator effects found in the dihydropyridine class of agents. These goals have been attained with isradipine. The pharmacological examination of the compound has continued to produce interesting and sometimes even unexpected results.

SOME GENERAL ASPECTS OF THE PHARMACOLOGY OF ISRADIPINE

It is not the purpose of this short review to examine in detail all aspects of the pharmacology of isradipine. A more complete review has appeared recently (13). Isradipine (under its code name PN 200-110) has become very well known in the scientific community because of its high affinity and high selectivity for the dihydropyridine binding site (8,14–17).

Like other calcium antagonists isradipine potently dilates blood vessels, especially the coronary artery *in vitro* (18,19) and *in vivo* (20). It also has cardioprotective effects even at non-cardiodepressant concentrations (21), when administered before the onset of ischaemia. Some of these protective effects were most likely attributable to the coronary vascular activity of isradipine, since it prevented the no-reflow phenomenon.

We have found already in the early stage of characterization that isradipine was a rather potent antihypertensive agent in spontaneously hypertensive rats, but with a surprisingly flat dose–response curve (19). Despite the potent anti-hypertensive action, isradipine caused rather pronounced diuresis in conscious rats (19). Furthermore, isradipine was recognized early as an unusually potent calcium antagonist on cerebral vessels *in vitro* (22) and *in vivo* (20).

ANTIVASOCONSTRICTOR EFFECTS IN NORMAL AND ATHEROSCLEROTIC RABBITS

Isradipine was found to possess very potent antivasoconstrictor effects against noradrenaline, phenylephrine and angiotensin II (23). All these effects are probably related to the action of isradipine on L-type calcium channels and there is no need to postulate effects on receptor operated channels (23). Interestingly, the antivasoconstrictor effects of isradipine against noradrenaline and phenylephrine were found to be enhanced in atherosclerotic rabbits compared with normal animals kept for the same period of time under the same conditions as the ones made atherosclerotic, except for the cholesterol content of the food (24). After the end of the experiments, the aorta of these animals was examined *in vitro*. It lacked the normal relaxant response to acetylcholine thus showing a defective endothelial function. Isradipine therefore appears to be capable to compensate some of the defective endothelial function of these animals.

These findings underscore the suitability of isradipine in the therapy of patients with an atherosclerotic vascular system. The prophylactic antiatherosclerotic effects of isradipine in different animal models such as the cholesterol fed rabbit (25) and the rat with balloon induced carotid lesions (26) have been reviewed (27,28) and described in detail by Dr Weinstein (p. 71–84).

ISRADIPINE IN HEART FAILURE

One of the important goals of our drug design program was the separation of the negative chronotropic from the negative inotropic effects of our derivatives. Such a selectivity has obvious advantages. In this respect PN 200-110 was one of the best compounds, even though other considerations (antihypertensive action) also influenced its selection (compare 29 with 18, 30, 31). The absence of cardiodepression was confirmed *in vivo* under experimental conditions which sensitize animals to cardiodepressant drug effects. One such condition is pretreatment with beta-blocking agents. There were clear differences between the haemodynamic effects of verapamil and isradipine after pretreatment of conscious rabbits with two different beta-adrenoceptor blockers. Furthermore, isradipine, in contrast to verapamil, did not inhibit AV conduction (32). Finally, we imitated certain aspects of autonomic dysfunction in heart failure in an open chest rabbit model (33). These animals were pretreated with propranolol, which eliminates the beta-adrenoceptor mediated inotropic effects of sympathetic stimulation on the heart. However, the vascular periphery remains under the influence of the alpha-adrenergic vasoconstrictor component of the sympathetic system. The cervical vagal nerves were cut, imitating the loss of parasympathetic activity observed in patients with heart failure. Finally, we measured myocardial contractile force with a strain gauge sewn onto the myocardium. This parameter is much less sensitive than other parameters of myocardial function to changes in haemodynamics elicited by peripheral vasodilatation. Under these experimental conditions both nifedipine and diltiazem had clear, dose-related cardiodepressant activity *in vivo*, whereas isradipine did not interfere with cardiac contractile force and increased cardiac output dose-dependently over the whole dose-range (33).

Isradipine was aimed at the indication hypertension, so the vasodilator properties were an important selection criterion in addition to the ones discussed above, and in this respect isradipine also excelled. A combination of pharmacological effects which include strong peripheral vasodilation (afterload reduction), absence of cardiodepression, and diuretic properties could be attractive for the treatment of heart failure. The therapeutic options in this indication are still limited. Calcium antagonists could offer advantages for certain forms of heart failure, especially when coronary heart disease or hypertension are contributing factors. Unfortunately, clinical experience is rather negative for those calcium antagonists investigated in this indication. The main question is, whether this as a class effect is true for all calcium antagonists; whether 'the bridge is *always* too far' (34), or whether the wrong calcium antagonists have been tried, perhaps even in the wrong patients under the wrong circumstances.

The wrong calcium antagonists

As described above calcium antagonists are all cardiodepressant but the relationship between myocardial effects and other actions may vary between drugs. In pharmacological experiments isradipine is clearly less cardiodepressant than other calcium antagonists for any degree of vasodilatation achieved (see 33 for details). The first generation of calcium antagonists may thus be less suited for the treatment of heart failure than isradipine.

The wrong patients

It is well established that not all patients with essential hypertension respond well to calcium antagonists, that there are non-responders. Unfortunately, little information appears to exist in this respect for heart failure. It appears that patients with advanced chronic congestive heart failure deteriorate with nifedipine (35). There are most likely other types of non-responders, but it is unknown, how to select patients that profit most from calcium antagonist therapy.

The wrong circumstances

Most patients with a failing heart are on diuretic therapy and have an activated renin-angiotensin system. Observations on normotensive and hypertensive patients indicate that under these circumstances calcium antagonists might loose part of their blood pressure lowering activity (36–40). Also, experimental observations in rats (41) and rabbits (42) indicate that calcium antagonists and diuretics do not interact in a favourable manner. Patients in chronic heart failure with high plasma renin activity did not respond well or even showed deterioration to nifedipine (35,43). It should be remembered that some calcium antagonists have quite pronounced diuretic properties themselves and hence, like diuretics, also tend to increase plasma renin activity (44). It could therefore be reasoned that side-effects related to activation of the renin system might be cumulative, when diuretics and calcium antagonists are used concomitantly.

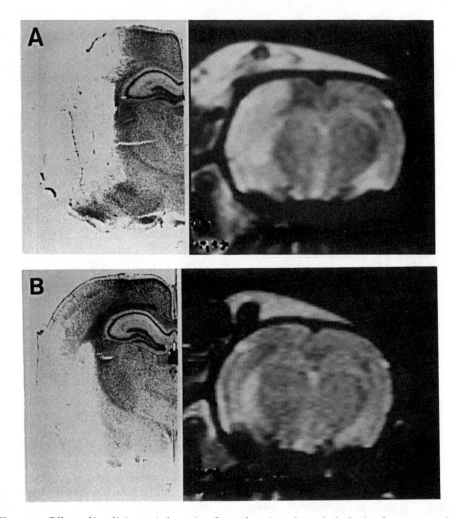

Figure 1 *Effects of isradipine on infarct size. Coronal sections through the brain of a representative control (A) and an isradipine (2.5 mg/kg s.c.) treated rat (B), made in vivo by magnetic resonance imaging, 24 hours following permanent occlusion of the left middle cerebral artery (right part of both figures). For validation, matching cryostatic sections were cut post-mortem and stained with cresyl violet. They are shown in the left part of the figures.*

Due to their anti-ischaemic activity calcium antagonists might be attractive especially in the treatment of ischaemic cardiomyopathy. Moreover, microvascular spasms have been suspected to play a pathogenetic role (45) in some forms of cardiomyopathies. If so, these might benefit most from calcium antagonist treatment. Patients in early heart failure might also profit because calcium antagonists can cause regression of left ventricular hypertrophy under certain circumstances in animals as well as in man (46–49). Isradipine with its pronounced vasodilator effect and its lack of cardiodepressant action might, unlike other calcium antagonists, be useful as an afterload reducing agent. If this drug retains its diuretic action (50) also in patients in heart failure, then isradipine could be useful as a single agent, perhaps especially in patients in the early stages of heart failure.

ISRADIPINE IN CEREBROVASCULAR DISEASE

Under the heading of general pharmacology the effects of isradipine on cerebral vessels *in vitro* and *in vivo* are already outlined (20,22). More recently, investigations related to cerebral ischaemia have been carried out and the results

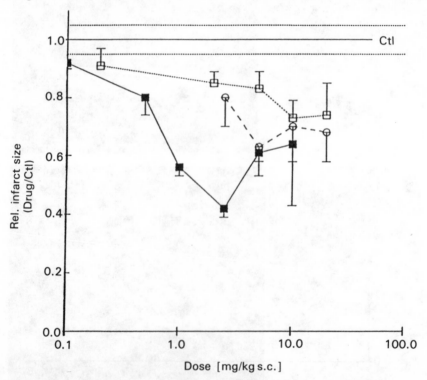

Figure 2 *Effects of calcium antagonists on infarct size: dose–response. Isradipine (filled squares), nimodipine (open circles) and darodipine (open squares) were injected s.c. at different doses as indicated, once immediately after permanent occlusion of the left middle cerebral artery. 24 hours later, the infarct size was determined from 4 coronal magnetic resonance images. The mean infarct size in vehicle injected control rats (CTL) was 18 000 voxels. The infarct size of drug treated animals is expressed as fraction of the infarct size of control animals. Means +/− SEM, number of animals per dose 6–10.*

are indeed encouraging. Firstly, dose–response curves obtained in anaesthetized cats showed that cerebral vascular effects were detectable at doses below those that caused systemic vascular effects, i.e. cerebral vasodilatation was seen at doses having only marginal systemic vasodilator effects. Furthermore the effects of isradipine on cerebral vessels lasted longer than those e.g. on coronary vessels or on blood pressure in both cats and rabbits (51).

Evidence has been obtained that isradipine possesses also flow-independent, cytoprotective effects (51–54). Detailed experiments were carried out recently in a stroke model based on permanent middle cerebral artery occlusion in rats. These rats were allowed to recover from anaesthesia and the infarct size was measured *in vivo* with nuclear magnetic resonance imaging as well as post-mortem, by standard histological methods. In this model isradipine dose-dependently reduced infarct size by up to 50–60% at the optimal dose as is shown in Fig. 1 and Fig. 2. Nimodipine, nitrendipine, darodipine and nicardipine were all clearly less effective, achieving about 40, 30, 20 and 10% reduction of infarct size respectively at the optimally active dose. The maximal efficacy of isradipine was obtained with a subcutaneous dose of 2.5 mg/kg administered at the time of occlusion (51). Interestingly, and in contrast to experiments on myocardial ischaemic damage, pretreatment was not necessary and even starting treatment after the onset of the cerebral infarction protected the ischaemic tissue (Fig. 3). As expected the effect was reduced with time and had vanished when treatment was initiated later than 4 hours after the onset of ischaemia (53).

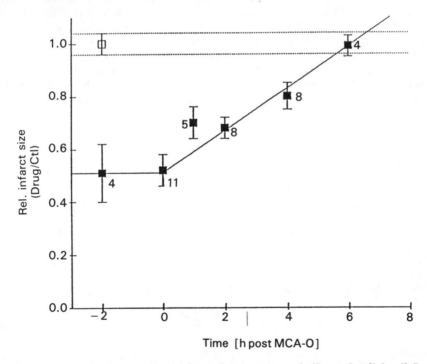

Figure 3 *Relationship between time of drug administration and efficacy. Isradipine (2.5 mg/kg/ was injected s.c. once, at different time points, as indicated, before and after permanent occlusion of the left middle cerebral artery. The infarct size (I) was determined from 4 coronal magnetic resonance images, 24 hours after stroke onset. The infarct size of drug treated animals is expressed as fraction of the infarct size of the control animals. Means +/− SEM, number of animals per point as indicated.*

Isradipine protected the brain also in an experimental situation which more closely imitated the normal use of an antihypertensive agent (53). Spontaneously hypertensive rats were treated for 6 days with one single dose of 2.5, 5 and 10 mg/kg isradipine per day. Blood pressure measured 12 hours after these doses was reduced by this treatment from about 200 to about 150, 135 and 120 mmHg respectively (it should be noted that these doses appear to be high, but it must be considered that rats metabolize isradipine rapidly, the half-life being in the order of 1 hour). The occlusion of the middle cerebral artery was also performed exactly 12 hours after the last dose and no further isradipine was administered after the occlusion. Nuclear magnetic resonance imaging again showed that infarct size decreased dose-dependently by 20, 40 and 60% after the 2.5, 5 and 10 mg/kg pretreatment respectively. The degree of protection was verified by histological determination of infarct size 5 days after the occlusion. These experiments show that the decrease of blood pressure achieved at the time of occlusion did not diminish the therapeutic efficacy of isradipine. In fact, the dose that normalized blood pressure also resulted in optimal protection against cerebral ischaemia.

In summary then, animal experiments indicate that isradipine displays interesting target tissue selectivities. Some of these have been confirmed in man during the development as an antihypertensive agent. Careful comparative studies covering wide dose-ranges in order to determine optimal efficacy, show pharmacodynamic properties which promise therapeutic benefits also in new indications, where so far calcium antagonists have not been used with great success. Areas where appropriate clinical studies should be undertaken in the future include chronic congestive heart failure and stroke. The preventive effect of isradipine on the progression of experimental atherosclerosis, again at doses similar to or even smaller than the ones used to elicit the desired haemodynamic effects is discussed in the contribution of Dr Weinstein. Experimental evidence suggests that the haemodynamic efficacy of isradipine is not diminished but much rather enhanced by atherosclerotic changes in the vasculature. The most important new findings, however, relate to the cerebral vascular and metabolic protection afforded.

REFERENCES

(1) Henry PD. Comparative pharmacology of calcium antagonists: nifedipine, verapamil and diltiazem. *Am J Cardiol* 1980; **46**: 1047–58.

(2) Henry PD. Chronotropic and inotropic effects of vasodilators. *Excerpta Medica ICS* 1979; **474**: 14–23.

(3) Perez JE, Borda L, Schuchleib R, Henry PD. Inotropic and chronotropic effects of vasodilators. *J Pharmacol Exp Ther* 1982; **221**: 609–13.

(4) Taira N. Differences in cardiovascular profile among calcium antagonists. *Am J Cardiol* 1987; **59**: 24B–9B.

(5) Reuter H. A variety of calcium channels. *Nature* 1985; **316**: 391.

(6) Nowycky MC, Fox AP, Tsien RW. Three types of neuronal calcium channel with different calcium agonist sensitivity. *Nature* 1985; **316**: 440–3.

(7) Murphy KMM, Gould RJ, Largent BL, Snyder SH. A unitary mechanism of calcium antagonist drug action. *Proc Natl Acad Sci* 1983; **80**: 860–4.

(8) Galizzi JP, Borsotto M, Barhanin J, Fosset M, Lazdunski M. Characterization and photoaffinity labeling of receptor sites for the Ca^{2+} channel inhibitors d-cis-diltiazem, ($\pm$)bepridil, desmethoxyverapamil, and (+)PN200-110 in skeletal muscle transverse tubule membranes. *J Biol Chem* 1986; **261**: 1393–97.

(9) Glossmann H, Ferry DR, Goll A, Rombush M. Molecular pharmacology of the calcium channel: evidence for subtypes, multiple drug-receptor sites, channel subunits, and the development of a radioiodinated 1,4-dihydropyridine calcium channel label, [^{125}I]Iodipine. *J Cardiovasc Pharmacol* 1984; **6** (Suppl 4): S608–21.

(10) Holck M, Osterrieder W. Inhibition of myocardial Ca^{2+} channels by three dihydropyridines with different structural features: potential-dependent blockade by Ro18-3981. *Br J Pharmacol* 1987; **91**: 61–7.

(11) Gähwiler BH, Brown DA. Effects of dihydropyridines on calcium currents in CA3 pyramidal cells in slice cultures of rat hippocampus. *Neuroscience* 1987; **20**: 731–8.

(12) Godfraind T, Morel N, Wibo M. Tissue specificity of dihydropyridine-type calcium antagonists in human isolated tissues. *Trends Pharmacol Sci* 1988; **9**: 37–9.

(13) Hof RP, Rüegg UT. Pharmacology of the new calcium antagonist isradipine and its metabolites. *Am J Med* 1988; **84** (Suppl 3B): 13–17.

(14) Ferry DR, Glossmann H. Identification of putative calcium channels in skeletal muscle microsomes. *FEBS Lett* 1982; **148**: 331–7.

(15) Lee HR, Roeske WR, Yamamura HI. High affinity specific [^{3}H]($+$)PN200-110 binding to dihydropyridine receptors associated with calcium channels in rat cerebral cortex and heart. *Life Sci* 1984; **35**: 721–32.

(16) Borsotto M, Barhanin J, Norman RI, Lazdunski M. Purification of the dihydropyridine receptor of the voltage dependent Ca^{2+} channel from skeletal muscle transverse tubules using ($+$)[^{3}H]PN200-110. *Biochem Biophys Res Comm* 1984; **122**: 1357–66.

(17) Cortes R, Supavilai P, Karobath M, Palacios JM. Calcium antagonist binding sites in the rat brain: quantitative autoradiographic mapping using the 1,4-dihydropyridines ^{3H}PN200-110 and ^{3H}PY108-068. *J Neural Transm* 1984; **60**: 169–97.

(18) Hof RP, Scholtysik G, Loutzenhiser R, Vuorela HJ, Neumann P. PN200-110, a new calcium antagonist: electrophysiological, inotropic and chronotropic effects on guinea-pig myocardial tissue and effects on contraction and calcium uptake of rabbit aorta. *J Cardiovasc Pharmacol* 1984; **6**: 399–406.

(19) Hof RP, Salzmann R, Siegl H. Selective effects of PN200-110 (isradipine) on the peripheral circulation and the heart. *Am J Cardiol* 1987; **59**: 30B–6B.

(20) Hof RP, Hof A, Scholtysik G, Menninger K. Effects of the new calcium antagonist PN200-110 on the myocardium and the regional peripheral circulation in anesthetized cats and dogs. *J Cardiovasc Pharmacol* 1984; **6**: 407–16.

(21) Cook NS, Hof RP. Cardioprotection by the calcium antagonist PN200-110 in the absence and presence of cardiodepression. *Br J Pharmacol* 1985; **86**: 181–9.

(22) Müller-Schweinitzer E, Neumann P. In vitro effects of the calcium antagonists PN 200-110, nifedipine, and nimodipine on human and canine cerebral arteries. *J Cereb Blood Flow Metabol* 1983; **3**: 354–61.

(23) Hof RP, Rüegg UT. Antivasoconstrictor effects of isradipine: a quantitative approach in anesthetized rats and conscious rabbits. *Am J Med* 1989; **86** (Suppl 4A): 50–6.

(24) Hof RP, Hof A. Vasoconstrictor and vasodilator effects in normal and atherosclerotic conscious rabbits. *Br J Pharmacol* 1988; **95**: 1075–80.

(25) Habib JB, Bossaller C, Wells S, Williams C, Morrisett JD, Henry PD. Preservation of endothelium-dependent vascular relaxation in cholesterol-fed rabbit by treatment with the calcium blocker PN200-110. *Circ Res* 1986; **58**: 305–9.

(26) Handley DA, VanValen RG, Melden MK, Saunders RN. Suppression of rat carotid lesion development by the calcium channel blocker PN200-110. *Am J Pathol* 1986; **124**: 88–93.

(27) Weinstein DB, Heider JG. Antiatherogenic properties of calcium antagonists. *Am J Cardiol* 1987; **59**: 163B–72B.

(28) Weinstein DB, Heider JG. Antiatherogenic properties of calcium channel blockers. *Am J Med* 1988; **84** (Suppl 3B); 102–8.

(29) Hof RP, Scholtysik G. Effects of the calcium antagonist PY108-068 on myocardial tissues *in vitro* and on reflex tachycardia *in vivo*. *J Cardiovasc Pharmacol* 1983; **5**: 176–83.

(30) Rüegg UT, Scholtysik G, Cook NS, Vogel A, Hof RP. PN200-110, a calcium antagonist with high potency and selectivity for vascular smooth muscle. *Ann NY Acad Sci* 1988; **522**: 304–5.

(31) Wada Y, Satoh K, Taira N. A study on the separation of the coronary vasodilator from cardiac effects of PN200-110, a new dihydropyridine calcium antagonist, in the dog heart. *J Cardiovasc Pharmacol* 1985; **7**: 190–6.

(32) Hof RP. Interaction between two calcium antagonists and two beta blockers in conscious rabbits: hemodynamic consequences of differing cardiodepressant properties. *Am J Cardiol* 1987; **59**: 43B–51B.

(33) Hof RP. Comparison of cardiodepressant and vasodilator effects of PN200-110 (isradipine), nifedipine and diltiazem in anesthetized rabbits. *Am J Cardiol* 1987; **59**: 37B–42B.

(34) Packer M, Kessler PD, Lee WH. Calcium-channel blockade in the management of severe chronic congestive heart failure: a bridge too far. *Circulation* 1987; 75 (Suppl V): V-56–V-64.

(35) Packer M, Lee WH, Medina N, Yushak M, Bernstein JL, Kessler PD. Prognostic importance of the immediate hemodynamic response to nifedipine in patients with severe left ventricular dysfunction. *J Am Coll Cardiol* 1987; **10**: 1303–11.

(36) MacGregor GA, Markandu ND, Smith SJ, Sagnella GA. Captopril: contrasting effects of adding hydrochlorothiazide, propranolol, or nifedipine. *J Cardiovasc Pharmacol* 1985; **7** (Suppl 1): S82–7.

(37) Resnick LM, Nicholson JP, Laragh JH. Calcium metabolism and renin–aldosterone system in essential hypertension. *J Cardiovasc Pharmacol* 1985; **7** (Suppl 6): S187–93.

(38) Bellini G, Battilana G, Puppis E *et al*. Renal responses to acute nifedipine administration in normotensive and hypertensive patients during normal and low sodium intake. *Curr Ther Res* 1984; **35**: 974–81.

(39) MacGregor GA, Markandu ND, Smith SJ, Sagnella GA. Does nifedipine reveal a functional abnormality of arteriolar smooth muscle cell in essential hypertension— the effect of altering sodium balance. *J Cardiovasc Pharmacol* 1985; **7** (Suppl 6): S178–81.

(40) Morgan T, Anderson A, Wilson D, Myers J, Murphy J, Nowson C. Paradoxical effect of sodium restriction on blood pressure in people on slow-channel calcium blocking drugs. *Lancet* 1986; **1**: 793.

(41) Waeber B, Nussberger J, Brunner HR. Does renin determine the blood pressure response to calcium entry blockers? *Hypertension* 1985; **7**: 223–7.

(42) Hof RP, Hof-Miyashita A. Different peripheral vasodilator effects of isradipine in sodium-loaded and sodium-depleted rabbits. *Gen Pharmacol* 1988; **19**: 243–7.

(43) Lefkowitz CA, Mow GW, Armstrong PW. A comparative evaluation of hemodynamic and neurohumoral effects of nitroglycerin and nifedipine in congestive heart failure. *Am J Cardiol* 1987; **59**: 59B–63B.

(44) Hof RP, Evenou JP, Hof-Miyashita A. Similar increase in circulating renin after equihypotensive doses of nitroprusside, dihydralazine or isradipine in conscious rabbits. *Eur J Pharmacol* 1987; **136**: 251–4.

(45) Sonnenblick EH, Fein F, Capasso JM, Factor SM. Microvascular spasm as a cause of cardiomyopathies and the calcium-blocking agent verapamil as potential primary therapy. *Am J Cardiol* 1985; **55(B)**: 179–84.

(46) Pfeffer MA, Pfeffer JM. Pharmacologic regression of cardiac hypertrophy in experimental hypertension. *J Cardiovasc Pharmacol* 1984; **6** (Suppl 6): S865–9.

(47) Tubau JF, Wikman-Coffelt J, Massie BM, Szlachcic J, Parmley WW, Sievers R, Henderson S. Diltiazem prevents hypertrophy progression, preserves systolic function, and normalises myocardial oxygen utilisation in the spontaneously hypertensive rat. *Cardiovasc Res* 1987; **21**: 606–14.

(48) Messerli FH, Schmieder RE, Nunez BD. Heterogeneous pathophysiology of essential hypertension: implications for therapy. *Am Heart J* 1988; **112**: 886–93.

(49) Grellet JP, Bonoron-Adele SM, Tariosse LJ, Besse PJ. Diltiazem and left ventricular hypertrophy in renovascular hypertensive rats. *Hypertension* 1988; **11**: 495–501.

(50) Krusell LR, Jespersen LT, Schmitz A, Thomsen K, Lederballe Pedersen O. Repetitive natriuresis and blood pressure. Long-term calcium entry blockade with isradipine. *Hypertension* 1987; **10**: 577–81.

(51) Sauter A, Rudin M, Wiederhold KH, Hof RP. Cerebrovascular, biochemical, and cytoprotective effects of isradipine in laboratory animals. *Am J Med* 1989; **86** (Suppl 4A): 134–46.

(52) Sauter A, Rudin M. Effects of calcium antagonists on high-energy phosphates in ischemic rat brain measured by ^{31}P NMR spectroscopy. *Magn Reson Med* 1987; **4**: 1–8.

(53) Sauter A, Rudin M. Treatment of hypertension with isradipine reduces infarct size following stroke in laboratory animals. *Am J Med* 1989; **86** (Suppl 4A): 130–3.

(54) Sauter A, Rudin M. Calcium antagonists reduce the extent of infarction in rat middle cerebral artery occlusion model as determined by quantitative magnetic resonance imaging. *Stroke* 1986; **17**: 1228–34.

Vascular actions of calcium antagonists

H. A. J. Struyker Boudier and M. W. J. Messing

Department of Pharmacology, University of Limburg, Maastricht, The Netherlands

INTRODUCTION

The primary haemodynamic mechanism underlying the antihypertensive effect of calcium antagonists is their vasodilatory action. Calcium antagonists block the entry of calcium into vascular smooth muscle cells. Recent research has provided detailed knowledge of the mechanisms of cellular calcium entry, including the voltage, the receptor agonist and the stretch-dependent mechanisms (1–3). However, less attention has been given thus far to the in-vivo relevance of these mechanisms in relation to the control of vascular resistance.

Vascular resistance is determined by a complex network of both in parallel and in series coupled vascular segments (Fig. 1). There is marked heterogeneity with respect to physiological behaviour and pharmacological sensitivity of these different segments. The purpose of this paper is to describe the sites of action of calcium antagonists in the different vascular segments.

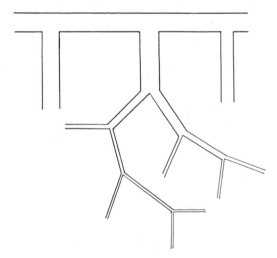

Figure 1. *The vascular system is composed of a network of parallel and in series coupled blood vessels.*

The use of isradipine and other calcium antagonists in cardiovascular diseases, edited by P. A. van Zwieten, 1989; Royal Society of Medicine Services International Congress and Symposium Series No. 157, published by Royal Society of Medicine Services Limited.

REGIONAL HAEMODYNAMICS OF CALCIUM ANTAGONISTS

With respect to the in parallel coupled vascular segments the question of regional selectivity of calcium antagonists is of relevance. Originally it was believed that calcium antagonists predominantly dilate the coronary vascular bed (4). Later, suggestions were made for other forms of regional selectivity for different dihydropyridines (5,6). With the introduction of sophisticated techniques for flow measurement, e.g. radioactive microspheres and implanted Doppler or electromagnetic flow probes, it has become possible to study the effects of calcium antagonists on simultaneously determined parallel vascular resistances. We have recently reviewed these experiments else-where (6). The major conclusions that can be drawn from all individual studies are:

—There is a marked difference in regional effects on vascular resistances for all calcium antagonists. The coronary and skeletal muscle vascular beds are dilated to the strongest degree. The skin is rather insensitive to the effects of calcium antagonists. Other vascular beds, e.g. brain, kidney and gastro-intestinal bed, have an intermediate sensitivity.

—The various calcium antagonists have a similar regional heterogeneity, even when compared in different species.

These two observations plead against vascular regional specificity as a result of differences in the calcium antagonist receptors in different vascular beds. Rather, the data suggest a heterogeneity on the basis of differences in local resistance control in various vascular beds. Recent studies point in the direction of myogenic tone as an important causal factor in the mode of vascular action of calcium antagonists (6,7). The coronary and skeletal muscle vascular beds have the highest degree of myogenic tone and these are exactly the tissues with highest sensitivity to the vasodilator effect of calcium antagonists. Ljung, Nordlander and their co-workers (7,8) have shown that the dihydropyridine derivative felodipine strongly inhibits myogenic tone. However, it is not certain whether all calcium antagonists inhibit myogenic tone to a similar degree (9).

Nievelstein et al. (10) have shown that the regional selectivity of calcium antagonist-induced vasodilatation disappears when nervous reflex control mechanisms are interrupted. In sino-aortic denervated or anaesthetized spontaneously hypertensive rats calcium antagonists of different classes cause a non-selective vasodilatation. These data suggest that in intact animals the regional selectivity of calcium antagonists is—at least partly—determined by a stronger influence of nervous reflex mechanisms on the kidney, gastro-intestinal bed and skin than on the coronary or skeletal muscle vascular bed. It would be of interest to investigate whether this pattern agrees with the relative roles of myogenic and nervous mechanisms in determining vascular tone in the different in parallel coupled vascular beds.

IN SERIES COUPLED VASCULAR SEGMENTS

Within one vascular bed different segments can be distinguished ranging from large arteries, arterioles, capillaries, venules to large veins. The concept of a precapillary sphincter as the primary site of resistance control should now be regarded as obsolete. Modern microcirculatory techniques have indicated the continuity of resistance control over the different segments. However, there is a marked heterogeneity with respect to the pharmacological sensitivity of the different vascular segments.

By far the most research on calcium antagonist-induced vasodilatation has thus far been performed on isolated large arteries, such as the aorta. The primary mechanism of CA-induced large artery dilatation is blockade of voltage-operated calcium channels. This conclusion is based upon a large series of experiments (1,6,11), showing that K$^+$-induced contractions are much more readily inhibited by calcium antagonists than noradrenaline-induced contractions. A different picture is obtained in smaller, resistance-sized arterioles. In these vessels calcium antagonists block K$^+$- and noradrenaline-induced contractions to an equal degree (1,6,11). The reasons for this difference between large and small arteries are not known, but may relate to a noradrenaline-induced depolarization of smooth muscle cells from resistance-sized vessels, thus causing activation of a voltage-dependent calcium gating mechanism.

Calcium-dependent vascular contractions have been described for both arterial and venous preparations. A higher concentration of calcium antagonists is needed to dilate precontracted veins (6,11). Administration of calcium antagonists into intact animals or humans also causes little venous effects.

MICROCIRCULATORY EFFECTS OF CALCIUM ANTAGONISTS

Little systematic research has been performed thus far on the effects of calcium antagonists—and other cardiovascular drugs—on the microcirculation. Using an isolated perfused skeletal muscle preparation in cats, Gustafsson et al. (12,13) recently showed a marked effect by calcium antagonists on precapillary vessels. This effect was primarily due to an interference with myogenic vascular reactivity. Basal vascular tone and vascular tone induced by stimulation of sympathetic nerves were both less sensitive to calcium antagonists than vascular tone induced by myogenic vascular reactivity.

Recently several authors have used intravital microscopy to study the effects of calcium antagonists on the microcirculation. Local (14) as well as intravenous (15) administration of verapamil dilates small arterioles of rat cremaster muscle. However, verapamil did not affect the mesenteric and pial vasculature (14). Similarly, the dihydropyridine derivative nimodipine has little effect on pial arteriolar diameters (16). The reason for these discrepancies may well be that (skeletal) muscle is more sensitive to the effects of calcium antagonists, as we discussed previously.

In a recent series of studies in our laboratory we used the chronically implanted dorsal skin muscle chamber in the spontaneously hypertensive rat to investigate the microvascular effects of felodipine, nifedipine and verapamil (see Fig. 2). All three calcium antagonists caused a dose-dependent fall in blood pressure and dilated the small arterioles, ranging in diameter from 150–10 μm. The vasodilatory effect was more pronounced in the most distally located vessels. In addition, calcium antagonists caused a marked inhibition of spontaneous oscillations in vascular diameters, the so-called vasomotion (15,17).

The relatively strong effect of calcium antagonists on the distal sections of the microcirculation has several important implications. From a mechanistic point of view these observations indirectly support the myogenic theory of vasodilator activity of calcium antagonists, since the contribution of myogenic tone increases the more distally one gets into the microcirculation. A second important implication is that these microvascular observations may explain oedema formation during clinical use of calcium antagonists. The pronounced precapillary vasodilatation causes an increase in capillary hydrostatic pressure and thus in net transcapillary fluid transfer.

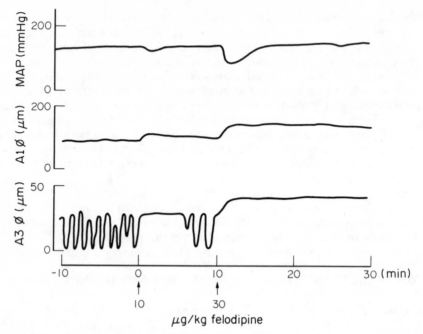

Figure 2. *The effects of intravenous administration of 2 different doses of felodipine on mean arterial pressure and diameters of an A1 and A3 arteriole in the dorsal backflap of the spontaneously hypertensive rat.*

LONG-TERM VASCULAR EFFECTS

The effectiveness of antihypertensive therapy may ultimately depend partly upon the degree to which a drug is able to reverse the alterations in structural design of the heart and blood vessels. Long-term therapy with calcium antagonists leads to a reduction of cardiac hypertrophy. The effects on vascular design are less clear with some authors reporting no and others a significant reduction of vascular hypertrophy (1). Calcium antagonists can inhibit vascular DNA synthesis, thereby reducing arterial smooth muscle cell proliferation (18).

Another interesting long-term vascular effect of calcium antagonists is that they may increase the amount of capillaries. Turek *et al.* (19) reported that treatment of spontaneously hypertensive rats with nifedipine leads to an increased cardiac capillary density. A similar effect has been reported for pial capillarization following nimodipine treatment (20).

CONCLUSIONS

Calcium antagonists lower blood pressure by dilating blood vessels. They do so with a preference for the coronary and skeletal muscle vascular bed. The vasodilatory effect is more prominent on arterial than on venous segments. Within the microcirculation the distal arterioles are the most sensitive to the effects of calcium antagonists. Both the regional and microcirculatory sensitivities to calcium antagonists support the concept of inhibition of myogenic tone as a primary mechanism of vascular action of calcium antagonists.

REFERENCES

(1) Struyker Boudier HAJ, Smits JFM, De Mey JGR. The pharmacology of calcium antagonists: a review. *J Cardiovasc Pharmacol* 1989; in press.

(2) Nayler WG. *Calcium antagonists.* London: Academic Press, 1988.

(3) Van Breemen C, Cauvin C, Johns A, Leijten P, Yamamoto H. Ca^{2+} regulation of vascular smooth muscle. *Fed Proc* 1986; 45: 2746–51.

(4) Fleckenstein A. *Calcium antagonism in heart and smooth muscle.* New York: Wiley, 1983.

(5) Hof RP. Selective effects of different calcium antagonists on the peripheral circulation. *Trends Pharmacol Sci* 1984; 5: 100–2.

(6) Struyker Boudier HAJ, De Mey JGR, Smits JFM, Nievelstein HMNW. Hemodynamic actions of calcium entry blockers. *Prog Basic Clin Pharmacol* 1989; 2: 21–67.

(7) Ljung B, Nordlander M, Johansson B. Felodipine sensitivity *in vivo* and *in vitro* of activation pathways in vascular smooth muscles. *J Cardiovasc Pharmacol* 1987; 10 (Suppl 1): 89–95.

(8) Nordlander M, Thalén P. Effects of felodipine on local and neurogenic control of vascular resistance. *J Cardiovasc Pharmacol* 1987; 10 (Suppl 1): 100–6.

(9) Hwa JJ, Bevan JA. A nimodipine-resistant calcium pathway is involved in myogenic tone in a resistance artery. *Am J Physiol* 1986; 251: H182–9.

(10) Nievelstein HMNW, Van Essen H, Tyssen CM, Smits JFM, Struyker Boudier HAJ. Systemic and regional actions of calcium entry blockers in conscious spontaneously hypertensive rats. *Eur J Pharmacol* 1985; 113: 187–98.

(11) Cauvin C, Loutzenhiser R, Van Breemen C. Mechanisms of calcium antagonist induced vasodilatation. *Ann Rev Pharmacol Ther* 1983; 23: 373–96.

(12) Gustafsson D, Grände PO, Borgström P, Lindberg L. Effects of calcium antagonists on myogenic and neurogenic control of resistance and capacitance vessels in cat skeletal muscle. *J Cardiovasc Pharmacol* 1988; 12: 413–22.

(13) Gustafsson D. Microvascular mechanisms involved in calcium antagonist edema formation. *J Cardiovasc Pharmacol* 1987; 10 (Suppl 1): 121–31.

(14) Altura BM, Altura BT, Gebrewold A. Differential effects of the calcium antagonist verapamil on lumen sizes of terminal arterioles and muscular venules in the rat mesenteric, pial and skeletal muscle microvasculatures. *Br J Pharmacol* 1980; 70: 351–3.

(15) De Clerck F, Loots W, Voeten J, Janssen PAJ. Differential effects of verapamil and flunarazine on epinephrine-induced vasoconstriction and on spontaneous vasomotion of arterioles in skeletal muscle in the rat *in vivo*. *J Cardiovasc Pharmacol* 1989; 13: 76–83.

(16) Haws CW, Heisted DD. Effects of nimodipine on cerebral vasoconstrictor responses. *Am J Physiol* 1984; 247: H170–6.

(17) Hundley WG, Renaldo GJ, Levasseur JE, Kontos HA. Vasomotion in cerebral microcirculation of awake rabbits. *Am J Physiol* 1988; 254: H67–71.

(18) Nilsson J, Sjölund M, Palmberg L, Van Euler AM, Jonzon B, Thyberg J. The calcium antagonist nifedipine inhibits arterial smooth muscle proliferation. *Atherosclerosis* 1985; 58: 109–22.

(19) Turek Z, Kubat K, Kazda S, Hoofd L, Rakusan K. Improved myocardial capillarisation in spontaneously hypertensive rats treated with nifedipine. *Cardiovasc Res* 1987; 21: 725–9.

(20) Yuan XQ, Prough DS, Dusseau JW *et al.* The long-term effects of nimodipine on pial microvasculature and systemic circulation in conscious rats. *Am J Physiol* 1989; in press.

Calcium antagonists in hypertension

A. J. Man in 't Veld

Department of Internal Medicine I, University Hospital Dijkzigt, Erasmus University Rotterdam, The Netherlands

SUMMARY

Calcium antagonists are effective, safe and well tolerated agents for the treatment of arterial hypertension. They cause few metabolic disturbances as compared with other antihypertensive drugs. Combination with diuretics, beta-blockers and angiotensin converting enzyme inhibitors results in additive effects on blood pressure. In hypertensive patients with diabetes mellitus, chronic obstructive lung disease, congestive heart failure, gout, renal failure, peripheral atherosclerotic disease or Raynaud's phenomena calcium antagonists are not contraindicated. Dietary sodium restriction during antihypertensive therapy with calcium antagonists is not required for optimal antihypertensive efficacy. Some of the second generation dihydropyridine type calcium antagonists, with greater potency, vascular selectivity and longer duration of action, will optimize treatment of hypertension with these agents. Their antiatherosclerotic, antiplatelet and 'antitrophic' effects in experimental models for atherogenesis and hypertension hold a promise for the future, since so far we have not been particularly successful in reducing the toll of coronary death by the treatment of hypertension.

INTRODUCTION

Originally the calcium antagonists were used extensively in the treatment of angina (1,2) and tachyarrhythmias (1,3,4). It was already in 1962 when it was noted that verapamil lowered blood pressure in hypertensive subjects, when given intravenously (5). It would take, however, almost 20 years before the clinical usefulness of these agents for the treatment of hypertension was fully recognized. Some reasons for this delay might have been the increasing popularity of the beta-adrenoceptor antagonist over this period of time, as well as the relatively poor bioavailability and short duration of action of verapamil and nifedipine as the only available calcium antagonists in this period. Furthermore, most anti-hypertensive regimens based on traditional vasodilators, like hydralazine, have been poorly tolerated by patients, because of the high incidence of undesirable side-effects, such as reflex stimulation of the sympathetic and renin angiotensin systems, which results in tachycardia, renal retention of sodium and water and the development of tachyphylaxis. The rising research effort to unravel the

The use of isradipine and other calcium antagonists in cardiovascular diseases, edited by P. A. van Zwieten, 1989; Royal Society of Medicine Services International Congress and Symposium Series No. 157, published by Royal Society of Medicine Services Limited.

intracellular ion abnormalities in hypertension has undoubtedly contributed to the change of interest in calcium antagonists in recent years.

Calcium antagonists have now emerged as an important 'new' class of drugs in the treatment of hypertension. A number of excellent reviews on their use in hypertension have appeared over the last five years (6–20). This review is an attempt to summarize and update the clinically relevant information on the current use of calcium antagonists in hypertension.

EFFICACY OF CALCIUM ANTAGONISTS IN HYPERTENSION

Comparative efficacy of monotherapy

The antihypertensive potential of the calcium antagonists is firmly established beyond any doubt. Calcium antagonists like verapamil, diltiazem, nifedipine and isradipine have now been extensively compared in double blind, controlled and otherwise well designed trials with beta-blockers, diuretics, angiotensin converting enzyme inhibitors, labetalol and prazosine (21–29). On average, the calcium antagonists are at least as effective in reducing both systolic and diastolic blood pressure as the other antihypertensive agents. Whereas on monotherapy with either beta-blockers, diuretics or angiotensin converting enzyme inhibitors responder-rates of 50 to 60% are normally found, unusually high responder-rates, up to 80%, have been found with some of the calcium antagonists, like isradipine (29).

Combination therapy

If monotherapy with a calcium antagonist fails to control blood pressure sufficiently the drug can be safely combined with beta-blockers, angiotensin converting enzyme inhibitors or diuretics (30–40). Whereas in most studies the addition of a beta-blocker or an angiotensin converting enzyme inhibitor to a calcium antagonist seems to result in simply additive effects on blood pressure this has been questioned to be the case for the diuretics (37,38). Part of the controversy might have arisen because of inadequacies in trial design, particularly the failure to take into account placebo responses, baseline blood pressure shifts and period effects, or the inclusion of too few subjects to give some trials sufficient power to demonstrate a significant fall in blood pressure. Furthermore, it is sometimes difficult to compare one trial with another with respect to age, race, severity of hypertension and renin status in the patient populations under study. These factors may all influence responsiveness to therapy with either class of agent (38). In a recent double blind cross-over comparison of nitrendipine and hydrochlorothiazide, however, it was clearly shown in 87 subjects over periods of 8 weeks that the two compounds had additive effects on blood pressure (39). Therefore addition of a calcium antagonist as a second step after starting treatment with a low dose of a diuretic should be considered for those patients failing to respond to a diuretic alone, specifically if for any reason a beta-blocker cannot be given.

Efficacy in special subgroups: blacks, elderly, low-renin

It is well known that beta-blockers may be sometimes less effective in black hypertensives, the elderly and patients with low-renin hypertension. Angiotensin converting enzyme inhibitors may also exhibit lower responder-rates in black patients with hypertension in whom plasma renin is often low (41–46). However, in contrast to previous experience with the beta-blockers and angiotensin converting enzyme inhibitors the calcium antagonist also appear to be markedly effective in the older hypertensive individual, in patients with low plasma renin and in black subjects (41–49). It should be born in mind, however, that the correlations between blood pressure response, plasma renin and age, although statistically significant, may have little predictive value for the individual patient. Correlation coefficients around and often below 0.5 for the relationships between the mentioned variables limit the practical value of these scientifically interesting facts for the tailoring of treatment in the hypertensive individual. Furthermore, some of the findings regarding the relationship between age and blood pressure response to calcium antagonists can be criticized on methodological grounds (49).

HAEMODYNAMICS OF CALCIUM ANTAGONISTS IN HYPERTENSION

Peripheral vascular resistance

The principal mechanism underlying the antihypertensive action of the calcium antagonists is the vasodilatation that results from interference with the excitation-contraction coupling in peripheral vessels (50–54). Consequently the haemodynamic abnormality in arterial hypertension, i.e. the raised total peripheral vascular resistance, is partly or wholly reversed. Initially the vasodilator effect of the calcium antagonists evokes a baroreflex mediated activation of the sympathetic and renin-angiotensin system. Increases in myocardial contractility, heart rate, plasma renin and noradrenaline are observed. In contrast with the classical vasodilators, like hydralazine, sodium and water retention does not occur and 'pseudo-tolerance' or tachyphylaxis are not observed. The potential mechanism underlying this phenomenon is discussed below. Depending on the type of calcium antagonist and its selectivity for cardiac versus vascular tissue, reflex cardiostimulation may not be observed, as is the case for drugs like verapamil and diltiazem.

Renal haemodynamics

Characterization of the renal effects of calcium antagonists has not been easy because they alter several regulatory functions within the kidney. Whereas the calcium antagonists have little effect in the vasodilated isolated kidney, they profoundly inhibit the response of the kidney to vasoconstrictor substances (55–59). As a consequence renal blood flow and glomerular filtration rate increase, an effect which seems to be more pronounced in hypertensive than in normotensive subjects (55).

Renal sodium excretion

The calcium antagonists differ from the traditional vasodilators in that they increase rather than decrease renal excretion of sodium and water (60–65). Several mechanisms could play a role in the natriuretic and diuretic action of the calcium antagonists: (1) a change in the glomerular filtration rate and/or renal blood flow; (2) interference with renin release; (3) interference with aldosteron release or the action of aldosteron on distal tubules; (4) interference with adrenergic sodium handling and (v) a direct effect on tubular sodium handling. Several of these mechanisms could play a role on the handling of sodium and water by the kidney during calcium channel blockade, although the last type of action seems to be the most important one (61). It has been shown, that felodipine increases renal blood flow without a change in glomerular filtration rate (61,62). However, the increase in renal blood flow often observed after several types of calcium antagonists is not necessarily associated with a natriuretic effect, and the natriuretic effect fades away during prolonged treatment with some of the calcium antagonists, whereas the increase in renal blood flow persists. The effects of different types of calcium antagonists on plasma renin and aldosterone are variable, although at least after acute administration of the dihydropyridine analogues renin and aldosterone tend to increase rather than decrease at a time that their natriuretic and diuretic action is most pronounced. The hypothesis of the interference of the calcium antagonists with the action of aldosterone on the distal tubule is interesting, but has not been tested in humans so far. Part of the action of the calcium antagonists may be neurally dependent, for instance through interference with alpha-receptor mediated mechanisms, although it should be taken into account that the drugs also exhibit these effects in the denervated kidney (60). The remaining and most likely mechanism is a direct inhibition of tubular sodium reabsorption by the calcium antagonists. It should be taken into account that the diuretic and natriuretic actions of calcium antagonists can be demonstrated in the first days of administration and that a negative sodium balance can be maintained for up to one week (60–62). Whether or not these effects of the compounds contribute to the onset and/or maintenance of their antihypertensive effect was an open question until recently. Krusel et al (63), recently reported the long-term (3.5 months) effects of one of the newer potent and vasculoselective dihydropyridines, isradipine, on renal haemodynamics and excretional parameters in subjects with essential hypertension. Even during long-term treatment, after the morning dose of isradipine blood pressure and renal vascular resistance were decreased and renal blood flow and glomerular filtration rate were increased. Output of fluid from the proximal tubules, as measured by the clearances of lithium and uric acid were increased and reabsorption of sodium beyond the proximal tubular level decreased. Consequently, sodium clearance and diuresis were increased. The changes in blood pressure were positively correlated with the changes in absolute proximal reabsorption of sodium, excretion of sodium and diuresis. The authors concluded that the natriuretic properties of the calcium antagonists may be more important for their long-term vasodilator and antihypertensive effect that previously assumed. This may be a particular feature for these newer potent and vasculoselective calcium antagonists and does not necessarily hold for other calcium antagonists as was shown by Marone et al (65) who found a small but significant sodium retention over the extracellular compartment during chronic treatment with nifedipine. This sodium retention could be reversed by addition of a thiazide diuretic, which also resulted in a further fall in blood pressure.

Calcium antagonists and dietary sodium intake

As mentioned earlier the usefulness of addition of diuretics to calcium antagonist has been a matter of debate (37–41, 65–70). Leonetti et al (69) showed that the acute hypotensive response after nifedipine was present both in the sodium deplete and sodium replate state, but that the fall in blood pressure was smaller after nifedipine during sodium restriction. Whereas the antihypertensive response to beta-blockers, diuretics and angiotensin converting enzyme inhibitors is clearly potentiated by dietary sodium restriction to below 80 mmols per day, this is not the case with the calcium antagonists verapamil (66), nifedipine and nitrendipine (70). Thus, dietary sodium restriction may not be necessary or appropriate in the treatment of essential hypertension with calcium antagonists. More importantly, the antihypertensive effect to calcium antagonists during a high-sodium intake is even not blunted.

HORMONAL AND METABOLIC EFFECTS OF CALCIUM ANTAGONISTS

Calcium is a component of many metabolic processes. By blocking calcium transport across cell membranes, calcium antagonists can therefore not only influence excitation-contraction coupling, but also excitation-secretion coupling. Accordingly, they could affect numerous metabolic and hormonal processes. Indeed, *in vitro* studies have often documented such effects, but these experiments are not necessarily relevant for the clinical situation where much lower plasma concentrations of the calcium antagonists are reached than in *in vitro* experiments.

Renin, angiotensin, aldosterone

It has been mentioned already, that the initial antihypertensive response of the dihydropyridines is often accompanied by a rise in plasma renin. This is generally not associated with a rise in plasma aldosterone. Review of the effects of diltiazem, nifedipine, verapamil and nitrendipine on the renin-angiotensin-aldosterone system shows that in general the long-term effects of these drugs on this system are negligible (71,72). Furthermore, the drugs have no effects on serum electrolytes, body composition, body weight and they exert a modest uricosuric action.

Glucose, calcium, the pituitary, catecholamines

Schoen *et al* reviewed the effects of the calcium antagonists on glucoregulatory, calcium regulatory, anterior and posterior pituitary hormones and plasma catecholamines (72). Despite the widespread involvement of calcium in the actions of hormones, it appears that the calcium antagonists have no dramatic impact on hormone regulation.

Plasma lipids

Calcium antagonist do not exhibit so-called unfavourable effects on plasma lipids (73,74). In contrast to diuretics and most beta-blockers these agents do not reduce HDL-cholesterol, and they do not increase LDL-cholesterol. The recent observation that isradipine even had a positive effect on plasma lipids in patients with essential hypertension, in that it increased the HDL-cholesterol in the apolipoprotein AI and AII subfractions after 13 weeks of treatment, is of interest and demands further studies. Review of 43 recent studies of several calcium antagonists on plasma lipids and lipoprotein subfractions showed that lipid metabolism is not affected by the majority of these drugs (74).

CALCIUM ANTAGONIST IN THE TREATMENT OF HYPERTENSION WITH CONCOMITTANT DISEASE

Diabetes mellitus

Early in vitro experiments with calcium antagonists have led to the assumption that a deterioration in carbohydrate metabolism is to be expected if patients with diabetes mellitus are being treated with these drugs. Review of 74 publications on glucose homeostasis in non-diabetics and 35 papers on long-term therapy of diabetics with calcium antagonists shows that the currently used calcium antagonist and the doses in which they are used, have no untoward effects on glucose metabolism (74). Thus, the calcium antagonists may be the drug of choice in patients with hypertension and diabetes in whom treatment with diuretics and some beta-blockers is relatively contra-indicated.

Bronchial asthma

Beta-blocking agents are contra-indicated for the hypertensive patient with asthma. Originally it was hoped that the calcium antagonists would also cause bronchodilatation in patients with chronic obstructive lung disease and bronchial asthma (75,76). Although these hopes have appeared to be premature it should be noted that the calcium antagonists do not negatively influence airway resistance, not even after histamine challenge. Therefore they can be safely used in subjects with obstructive pulmonary disease.

Congestive heart failure

Beta-blockers may precipitate heart failure as a consequence of their negative inotropic effect, although this appears not to be a major problem with these drugs in the hand of the careful physician. Despite some early promising reports of the effects of verapamil and nifedipine in patients with less severe forms of congestive heart failure, it was recognized fairly soon, that the negative inotropic actions of drugs like nifedipine, verapamil and diltiazem might be a threat for the patient with severe congestive heart failure (77,78). Initial experience with one of the newer, potent and vasculoselective dihydropyridine analogues isradipine are very promising in this context (79–81).

SIDE-EFFECTS OF CALCIUM ANTAGONISTS

The most common side-effects of calcium antagonists are a direct consequence of their actions on smooth muscle contraction and their relative non-selectivity for intestinal smooth muscle and certain vascular beds (Table 1, 82–86). It is therefore hoped that the development of newer calcium antagonists with greater selectivity for the vascular wall or even subselectivity for certain vascular beds will result in fewer side-effects. This seems to be the case for a drug like isradipine (84). Ankle oedema during treatment with the calcium antagonists is not a consequence of renal retention of sodium and water caused by their negative inotropic effect. Gustafsson (83) clearly showed in elegant experiments that the ankle oedema results from a transfer of fluid from the plasma compartment to the interstitial space driven by an increased capillary hydrostatic pressure due

Table 1 *Common side-effects of some calcium antagonists*

	Verapamil	Diltiazem	Nifedipine	Felodipine	Isradipine
Headache	8%	2%	6%	7%	3%
Flushing	8%	1%	12%	5%	2%
Nausea	1%	3%	4%	—	—
Constipation	30%	22%	—	—	—
Bradycardia	2%	1%	—	—	—
Ankle oedema	—	2%	15%	15%	5%
Tachycardia	—	—	15%	5%	4%

to the relative selectivity of the calcium antagonists for the arterial side of the circulation.

DRUG INTERACTIONS

Some clinically relevant drug interactions with the calcium antagonists are listed in Table 2. Particularly with verapamil plasma digoxine levels may rise to within the toxic range. This effect is less pronounced with the dihydropyridine compounds and it has been reported not to occur after isradipine (85). The potentiation of the antihypertensive effect of alpha-blockers can be used for therapeutic purposes as well as the inhibition of reflex tachycardia after the acute administration of some of the calcium antagonists by the beta-adreno-ceptor blocking agents. Intravenous administration of verapamil to a patient already treated with a beta-blocker should be avoided, since the negative ino-tropic and chronotropic effects of the drugs may be additive and serious con-duction and rythm disturbances may result. The effects of H_2-antagonists and drugs that increase the hepatic microsomal oxidase activity are to be remembered so that the dosage of the calcium antagonists can be adjusted in case of increased occurrence of side-effects or loss of antihypertensive efficacy, respectively.

FUTURE PERSPECTIVES OF CALCIUM ANTAGONIST THERAPY IN HYPERTENSION

Despite the availability of a wide range of antihypertensive agents, we have not been particularly successful in reducing the toll of atherosclerotic coronary artery disease in hypertension. In this context the calcium antagonists may hold some promises for the future.

Regression of cardiac and vascular hypertrophy

Cardiac hypertrophy as a complication of long-standing arterial hypertension represents an additional risk for the hypertensive individual, both in terms of morbidity and mortality. It is therefore of interest, that some of the dihydropyridine analogues have shown experimentally as well as in the clinical setting to reduce left ventricular hypertrophy (54,87,88). Drayer *et al*, however, did not find a significant decrease in left ventricular mass in patients with mild hypertension after a 12-month period of treatment with nitrendipine (89).

Table 2 *Drug interactions with calcium antagonists*

Digoxin	Increases digoxin levels, particularly with verapamil, not with isradipine
Alpha-blockers	Additive/synergistic effect on blood pressure
Beta-blockers	Reduced reflex-tachycardia (cave i.v. verapamil)
Anti-arrhythmics	Avoid combination
H_2-antagonists	Increases calcium antagonist levels
Drugs increasing hepatic micro- somal oxidase activity	Reduces calcium antagonist levels
Anaesthetics	Hypotension, bradycardia

Structural vascular changes in the precapillary arterioles are a cardinal feature of established hypertension. Since the calcium antagonists do not evoke reflex stimulation, of what are considered as 'trophic substances' (angiotensin and catecholamines) they may be of importance for the regression of structural vascular changes in hypertension (90). Although some recent animal work suggests that the calcium antagonists may be useful in this respect, there are few convincing clinical data available to date.

Antiatherogenic properties of calcium antagonists

Several calcium antagonists have antiatherogenic properties in experimental models (91–95). The relevance of these models for the clinical situation has been questioned. In addition it has been argued that the antiatherogenic effects of calcium antagonists can be only demonstrated in these models at concentrations of the drugs that are not attained in the clinical situation (91). Therefore it is important to note, that in the model of the cholesterol-fed rabbit isradipine inhibited cholesterol accumulation at doses that approximate the human dose (94). Studies show that although the calcium antagonists may protect against the development of experimental atherosclerosis, they are less effective in inducing its regression. So, it is clear that one must wait for the results of clinical studies. It should be realized that for the prevention, attenuation or reversal of the atherosclerotic process in the clinical setting, one faces the difficult methodological questions about the natural history of atherosclerosis in general and of coronary artery disease in particular and how to measure and to quantify it (96,97).

Effects of calcium antagonists on platelet function and thrombosis

The control of platelet function centres on the concentration of free intracellular calcium ions. An increase in the intracellular Ca^{2+} will result in platelet activation and eventually vasoconstriction, vascular damage, thrombosis and ischaemia. Calcium channel blocking agents have the ability to reduce the intracellular free Ca^{2+} availability. Although the drugs clearly have anti-platelet properties *in vitro*, these findings have been less consistent *in vivo* (98,99). It is of importance, however, that they clearly act synergistically with other platelet drugs, allowing lower doses to be used to achieve a satisfactory inhibitory effect on platelet function.

THE POSITION OF
CALCIUM ANTAGONISTS IN
THE TREATMENT OF HYPERTENSION

The management of arterial hypertension has dramatically changed over the last century, and more particularly over the last three decades (Table 3). When different types of agents (diuretics, beta-blockers, vasodilators, centrally acting drugs) became available clear instructions were needed for the prescribing physicians, how and when to use them, in which order and in what kind of combination. This led to the initial concept of 'standard step-care', later followed by 'individual step-care', as based on the individual characteristics of the patient (17). These step-care regimens did not take sufficiently into account that: (1) some of the drugs showed different dose-response relations for antihypertensive effects and side-effects; (2) despite taking into account individual characteristics of patients non-responder-rates up to 50 per cent during monotherapy are not uncommon and (3) some drug combinations were only needed because one drug was necessary to counteract the side-effects evoked by another one. Now with the availability of newer drugs like calcium antagonists and angiotensin converting enzyme inhibitors we have gradually changed to what is now known as 'liberal step-care' (17). This strategy simply implies the following: control blood pressure as

Table 3 *Evolution of treatment of hypertension*

Till 20th century	Phlebotomy
1930	Sedatives
1940	Sympathectomy
1945	Na-restriction/nitroprusside
1950	Ganglion blockers
1955	Reserpine
1960	Methyldopa/diuretics
1965	Beta-blockers
1970	Standard step-care
1980	Individual step-care
1990	Liberal step-care

Careless steps with stepless care

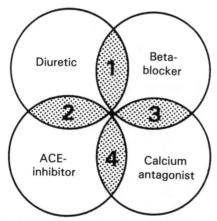

Figure 1 *Stepless care in the treatment of hypertension with fixed combination.*

effectively as possible, with as few as possible tablets, with as few as possible side-effects, metabolic derangements or negative interference with what is called 'quality of life'. How this goal eventually is achieved is less important than that it is achieved. In other words: it is the last step that is important and questions about the first choice in the 'first step' seem less relevant today, maybe that they are not even of academic interest. The calcium antagonists are effective, safe and well tolerated drugs that can be used in a wide variety of combinations with other antihypertensive agents. They can be used as first, intermediate or last choice in the management of hypertension as long as the criteria of 'liberal step-care' are fulfilled. As a consequence, it is to be expected, that in the near future therapies with fixed combinations will gain popularity (Fig. 1). Two of these combinations are already available, i.e. the beta-blocker-diuretic and the angiotensin converting enzyme inhibitor-diuretic combination. Undoubtedly the angiotensin converting enzyme inhibitor-calcium antagonist and beta-blocker calcium antagonist combinations will be added to these options. These combinations will result in a higher antihypertensive efficacy in the individual patient and a higher responder rate. Relatively low doses of the constitutions will cause a relatively low incidence of side-effects and good tolerability by the patient. Maybe that near the year 2000 the 'care of the individual steps' will have turned into optimal 'care of the individual patient': the era of 'careless steps with stepless care'.

REFERENCES

(1) Fleckenstein A. Fundamentale Herz- und Gefässwirkungen Ca^{2+}-antagonistische Koronartherapeutika. Med Klin 1975; 70: 1665–74.
(2) Fleckenstein-Grün G, Fleckenstein A. Calcium-Antagonismus, ein Grundprinzip der Vasodilatation. In: Fleckenstein A, Roskamm H, eds. Calcium-antagonismus. Berlin-Heidelberg-New York: Springer-Verlag, 1980; 91–102.
(3) Bender F. Die medikamentöse Behandlung von Herzrhythmusstörungen. Therapiewoche 1968; 42: 1803–07.
(4) Bender F. Die Behandlung der tachykarden Arrhythmien und der arteriellen Hypertonie mit Verapamil. Arzneim Forsch 1970; 20 (Suppl 9a): 1310–5.
(5) Heidland A, Klütsch K, Oebeck A. Myogenbedingte Vasodilatation bei Nierenischämie. Munch Med Wochenschr 1962; 35: 1636–7.
(6) Weidmann P, Gerber A, Laederach K. Calcium antagonists in hypertension. In: Advances in nephrology from the Necker Hospital. Year Book Medical Publishers 1984; 14: 197–232.
(7) Guazzi MD, Polese A, Fiorentini C, Bartorelli A, Moruzzi P. Treatment of hypertension with calcium antagonists. Review. Hypertension 1983; 5(Suppl II): II85–90.
(8) Klein W, Brandt K, Vrecko K, Härringer M. Role of calcium antagonists in the treatment of essential hypertension. Circ Res 1983; 57(Suppl 1): 174–81.
(9) Klein W. Treatment of hypertension with calcium antagonists: European data. Am J Med 1984; Oct 5: 143–8.
(10) Lederballe Pedersen O. Calcium blockade in arterial hypertension: Review. Hypertension 1983; 5(Suppl II): II74–79.
(11) Ram CVS. Southwestern Internal Medicine Conference: Calcium antagonists in the treatment of hypertension. Am J Med Sciences 1985; 290, 3: 118–32.
(12) Halparin AK, Cubeddu LX. The role of calcium channel blockers in the treatment of hypertension. Am Heart J 1985; 111, 2: 364–82.
(13) Robinson BF. Calcium-entry blocking agents in the treatment of systemic hypertension. Am J Cardiol 1985; 55: 102B–6B.
(14) Piepho RW. Individualisation of calcium entry-blocker dosage for systemic hypertension. Am J Cardiol 1985; 56: 105H–11H.

(15) Bühler FR, Bolli P, Erne P, Kiowski W, Müller FB, Hulthen UL, Hoa Ji B. Position of calcium antagonists in antihypertensive therapy. *J Cardiovasc Pharmacol* 1985; **7**(Suppl 4): S21–7.

(16) Müller FB, Bolli P, Erne P, Kiowski W, Bühler FR. Calcium antagonism—a new concept for treating essential hypertension. *Am J Cardiol* 1988; **57**: 50D–3D.

(17) Zanchetti A. Role of calcium antagonists in systemic hypertension. *Am J Cardiol* 1987; **59**: 130B–6B.

(18) Schoenberger JA. New approaches to a first-line treatment of hypertension. *Am J Med* 1988; **84**(Suppl 3B): 26–30.

(19) Fröhlich ED. Clinical pharmacology of calcium antagonists. *Hypertension* 1988; **11** (Suppl I): I222–4.

(20) Nayler W. *Calcium antagonists*. London: Academic Press, 1988: 101–11.

(21) Gould BA, Hornung RS, Mann S, Bala Subramanian V, Raftery EB. Nifedipine or verapamil as sole treatment of hypertension. An intra-arterial study. *Hypertension* 1983; **5**(Suppl II): II91–6.

(22) Rosei EA, Muiesan ML, Romanelli G, Castellano M, Beschi M, Corea L, Muiesan G. Similarities and differences in the antihypertensive effect of two calcium antagonist drugs verapamil and nifedipine. *J Am Coll Cardiol* 1986; **7**: 916–24.

(23) Doyle AE. Comparison of beta-adrenoceptor blockers and calcium antagonists in hypertension. *Hypertension* 1983; **5**(Suppl II): II103–8.

(24) Massie BM. Antihypertensive therapy with calcium-channel blockers: comparison with beta-blockers. *Am J Cardiol* 1985; **56**: 97H–100H.

(25) Massie BM, Tubau JF, Szlachcic J. Comparative studies of calcium channel blockers and beta-blockers in essential hypertension; clinical implications. *Circulation* 1987; **75**(Suppl V): V-163–9.

(26) Sowers JR, Mohanty PK, Comparison of calcium-entry blockers and diuretics in the treatment of hypertensive patients. *Circulation* 1987; **75**(Suppl V): V-170–3.

(27) Holzgreve H, Osterkom K, Runge J. Captopril in combination with hydrochloro-thiazide: comparative efficacy *vs* perceived best therapy. *Br J Clin Pharmac* 1987; **23**: 93S–101S.

(28) Fitzsimons TJ. Calcium antagonists: a review of the recent comparative trials. *J Hypertension* 1987; **5**(Suppl 3): S11–5.

(29) Kirkendall W. Comparative assessment of first-line agents for treatment of hyper-tension. *Am J Med* 1988; **84**(Suppl 3b): 32–41.

(30) Yagil Y, Kobrin I, Stessman J, Shanem J, Leibel B, Ben-Ishay D. Effectiveness of combined nifedipine and propranolol treatment of hypertension. *Hypertension* 1983; **5**(Suppl II): II113–7.

(31) Raftery EB. Calcium blockers and beta-blockers: alone and in combination. *Acta Med Scand* 1984; **694**(Suppl): 188–96.

(32) Ekelund LG. Nifedipine in combination therapy for chronic hypertension. *Am J Med* 1985; **79**(Suppl 4A): 41–3.

(33) Hansson L, Dahlöf B. Antihypertensive effect of a new dihydropyridine calcium antagonist, PN 200–110 (isradipine), combined with pindolol. *Am J Cardiol* 1987; **59**: 137B–40B.

(34) Dahlöf B, Andren L, Eggertsen R, Jern S, Svensson A, Hansson L. Long-term experience with the combination of pindolol and isradipine in essential hypertension. *Am J Med* 1988; **84**(Suppl 3B): 4–7.

(35) Brouwer RML, Bolli P, Erne P, Conen D, Kiowski W, Bühler FR. Antihypertensive treatment using calcium antagonists in combination with captopril rather than diuretics. *J Cardiovasc Pharmacol* 1985; **7**: S88–91.

(36) Guazzi MD, DeCesare N, Galli C, Salvioni A, Tramontana C, Tamborini G, Bartorelli A. Calcium channel blockade with nifedipine and angiotensin converting enzyme inhibition with captopril in the therapy of patients with severe primary hypertension. *Circulation* 1984; **2**: 279–84.

(37) MacGregor GA, Markandu ND, Smith SJ, Sagnella CA. Captopril: contrasting effects of adding hydrochlorothiazide, propranolol or nifedipine. *J Cardiovasc Pharmacol* 1985; **7**: S82–7.

(38) Sever PS, Poulter NR. Calcium antagonists and diuretics as combined therapy. *J Hypertension* 1987; **5**(Suppl 4): S123–6.

(39) Schoenberger JA. Calcium antagonists: use in hypertension evaluation of calcium antagonists in combination with diuretics. *Angiology* 1988; **January**: 87–93.

(40) Huhges GS, Cowart TD, Conradi EC. Efficacy of verapamil-hydrochlorothiazide-spironolactone therapy in hypertensive black patients. *Clin Pharmacy* 1987; **6**: 322–6.

(41) Bühler FR, Age and cardiovascular response adaptation. Determinants of an antihypertensive treatment concept primarily based on beta-blockers and calcium-entry blockers. *Hypertension* 1983; (Suppl III): III94–III100.

(42) Kiowski W, Bühler FR, Fadayomi MO, Erne P, Müller FB, Hulten UL, Bolli P. Age, race, blood pressure and renin: predictors for antihypertensive treatment with calcium antagonists. *Am J Cardiol* 1985; **56**: 81H–5H.

(43) Resnick LM, Laragh JH. Renin, calcium metabolism and the pathophysiologic basis of antihypertensive therapy. *Am J Cardiol* 1985; **56**: 68H–74H.

(44) Resnick L. Calcium metabolism, renin activity and the antihypertensive effects of calcium channel blockade. *Am J Med* 1986; **81**(Suppl 6A): 6–14.

(45) Resnick LM, Nicholson JP, Laragh JH. Calcium, the renin aldosterone system and the hypotensive response to nifedipine. *Hypertension* 1987; **10**: 254–8.

(46) Laragh JH. Calcium antagonists as a new treatment modality in hypertension. *Angiology* 1988; **January**: 100–5.

(47) Rowe JH. Approach to the treatment of hypertension in older patients. *Am J Med* 1988; **84**(Suppl 3B): 46–50.

(48) Giles TD, Massie BM. Role of calcium antagonists as initial pharmacologic monotherapy for systemic hypertension in patients over 60 years of age. *Am J Cardiol* 1988; **61**: 13H–17H.

(49) Chalmers JP, Smith SA, Wing LMH. Hypertension in the elderly: the role of calcium channel blocking drugs. *J Cardiovasc Pharmacol* 1988; in press.

(50) Lehmann HU, Hochrein H, Witte E, Mies SHW. Hemodynamic effects of calcium antagonists. *Hypertension* 1983; **5**(Suppl II): II66–73.

(51) Bolli P, Kiowski W, Erne P, Hulthen LU, Bühler FR. Hemodynamic and antihypertensive treatment responses with calcium antagonists. *J Cardiovasc Pharmacol* 1985; **7**: S126–30.

(52) Cody RJ. The hemodynamics of calcium channel blockers in hypertension: vascular and myocardial responses. *Circulation* 1987; **75**(Suppl I): I175–9.

(53) Lund-Johansen P, Omvik P, Central hemodynamic changes of calcium antagonists at rest and during exercise in essential hypertension. *J Cardiovasc Pharmacol* 1987; **10**(Suppl I): S139–48.

(54) Messerli FH, Oren S, Grossman E. Effects of calcium channel blockers on systemic hemodynamics in hypertension. *Am J Med* 1988; **84**(Suppl 3B): 8–12.

(55) Romero JC, Ray L, Granger JP, Ruilope LM, Rodicio JL. Multiple effects of calcium entry blockers on renal function in hypertension. *Hypertension* 1987; **10**: 140–51.

(56) Loutzenhiser R, Epstein M. Calcium antagonists and the kidney. *Hosp Practice* 1987; **January**: 63–76.

(57) Loutzenhiser RD, Epstein M. Renal hemodynamic effects of calcium antagonists. *Am J Med* 1987; **82**(Suppl 3B): 23–8.

(58) Bell PD. Calcium antagonists and intrarenal regulation of glomerular filtration rate. *Am J Nephrol* 1987; **7**(Suppl I): 24–31.

(59) Loutzenhiser R, Epstein M. Modification of the renal hemodynamic response to vasoconstrictors by calcium antagonists. *Am J Nephrol* 1987; **7**(Suppl I): 7–16.

(60) Zanchetti A, Stella A, Golin R, Adrenergic sodium handling and the natriuretic action of calcium antagonists. *J Cardiovasc Pharmacol* 1985; **7**(Suppl 6): S194–8.

(61) Zanchetti A, Leonetti G. Natriuretic effects of calcium antagonists *J Cardiovasc Pharmacol* 1985; **7**(Suppl 4): S33–7.

(62) Zanchetti H, Leonetti G. Discussion on the natriuretic effects of calcium antagonists. *J Cardiovasc Pharmacol* 1987; **10**(Suppl I): S161–4.

(63) Krusell LR, Jespersen LT, Schmitz A, Thomson K, Lederballe Pedersen O. Repetitive natriuresis and blood pressure. Long-term calcium entry blockade with isradipine. *Hypertension* 1987; **10**: 577–81.

(64) Luft FC. Calcium channel blocking drugs and renal sodium excretion. *Am J Nephrol* 1987; **7**(Suppl I): 39–43.

(65) Marone C, Luisoli S, Banio F, Beretta-Piccoli C, Bianchetti MG, Weidman P. Body sodium-volume state, aldosterone and cardiovascular responsiveness after calcium entry blockade with nifedipine. *Kidney Int* 1985; **28**: 658–65.

(66) Nicholson JP, Resnick LM, Laragh JH. The antihypertensive effect of verapamil at extremes of dietary sodium intake. *Ann Int Med* 1987; **107**: 329–34.

(67) MacGregor GA, Pevahouse JB, Cappuccio FP, Markandu ND, Nifedipine, diuretics and sodium balance. *J Hypertension* 1987; **5**(Suppl 4): S127–31.

(68) MacGregor GA, Pevahouse JB, Cappiccio FP, Markandu ND. Nifedipine, sodium intake, diuretics and sodium balance. *Am J Nephr* 1987; **7**(Suppl I): 44–8.

(69) Leonetti G, Rupoli L, Gradink R, Zanchetti A. Effects of a low-sodium diet on antihypertensive and natriuretic responses to acute administration of nifedipine. *J Hypertension* 1987; **5**(Suppl 4): S57–60.

(70) Luft FC, Weinberger MH. Review of salt restriction and the response to antihypertensive treatment. Satellite symposium on calcium antagonists. *Hypertension* 1988; **11**(Suppl I): I229–32.

(71) Bauer JH, Sunderrajan S, Reams G. Effects of calcium entry blockers on renin-angiotensin aldosterone system, renal function and hemodynamics, salt and water excretion and body fluid composition. *Am J Cardiol* 1985; **56**: 62H–7H.

(72) Schoen RE, Frishman WH, Shamoon H. Hormonal and metabolic effects of calcium channel antagonists in man. *Am J Med* 1988; **84**: 492–504.

(73) Rauramaa R, Tashinen E, Seppanen K, Rissanen V, Salonen R, Venalainen JM, Salonen JT. Effects of calcium antagonist treatment on blood pressure, lipoproteins and prostaglandins. *Am J Med* 1988; **84**(Suppl 3B): 93–6.

(74) Trost BN, Weidmann P. Effects of calcium antagonists on glucose homeostatis and serum lipids in non-diabetic and diabetic subjects: a review. *J Hypertension* 1987; **5**(Suppl 4): S81–104.

(75) Massey KL, Hendeles L. Calcium antagonists in the management of asthma. breakthrough or ballyhoo? *Drug Intell Clin Pharmacy* 1987; **21**: 505–9.

(76) Hendeles L, Harman E. Should we abandon the notion that calcium channel blockers are potentially useful for asthma? *J Allergy Clin Immunol* 1987; **79**, **6**: 853–5.

(77) Lefkowitz CA, Moe GW, Armstrong PW. Calcium antagonists: new therapy for congestive heart failure. *Chest* 1987; **91**, **1**: 1–2.

(78) Packer M, Kessler PD, Lee WH. Calcium channel blockade in the management of severe chronic heart failure: a bridge too far. *Circulation* 1987; **75**(Suppl V): V-56–64.

(79) Greenberg B, Deirdre S, Brondy D. Hemodynamic effects of PN 200–110 (Isradipine) in congestive heart failure. *Am J Cardiol* 1987; **59**: 70B–4B.

(80) O'Rouke RA, Walsh RA. Experience with calcium antagonist drugs in congestive heart failure. *Am J Cardiol* 1987; **59**: 64B–9B.

(81) Colucci WS. Usefulness of calcium antagonists for congestive heart failure. *Am J Cardiol* 1987; **59**: 52B–8B.

(82) Krebs R. Adverse reactions with calcium antagonists. *Hypertension* 1983; **5**(Suppl II): II125–9.

(83) Gustaffson D. Microvascular mechanisms involved in calcium antagonist edema formation. *J Cardiovasc Pharmacol* 1987; **10**(Suppl I): S121–31.

(84) Hamilton B. Treatment of essential hypertension with PN 200–110 (Isradipine). *Am J Cardiol* 1987; **59**: 141B–5B.

(85) Johnson BF, Wilson J, Marwaka R, Hock K, Johnson J. The comparative effects of verapamil and a new dihydropyridine calcium channel blocker on digoxin pharmaco-kinetics.

(86) Nayler W. *Calcium antagonists*. London: Academic Press, 1988: 261–81.

(87) Motz W, Strauer BE. Nifedipine in the long-term management of hypertensive heart disease. *Hypertension* 1983; **5**(Suppl II): II39–44.

(88) Kobayaski K, Tarazi RC. Effect of nitrendipine on coronary flow and ventricular hypertrophy in hypertension. *Hypertension* 1983; **5**(Suppl II): II45–51.

(89) Drayer JIM, Hall D, Smith VE, Weber MA, Wollam BL, White WB. Effect of the calcium channel blocker nitrendipine on left ventricular mass in patients with hypertension. *Clin Pharmacol Ther* 1983; **40**: 679–85.

(90) Hansson L. Regression of structural alterations of hypertension with calcium antagonists—vascular hypertrophy. *J Hypertension* 1987; **5**(Suppl 4): S71–4.

(91) Chobanian A. Effects of calcium channel antagonists and other antihypertensive drugs on atherogenesis. *J Hypertension* 1987; **5**(Suppl 4): S43–8.

(92) Parmley WW. Calcium channel blockers and atherogenesis. *Am J Med* 1987; **82**(Suppl 3B): 3–8.

(93) Weinstein DB, Heider JG. Antiatherogenic properties of calcium antagonists. *Am J Cardiol* 1987; **59**: 163B–72B.

(94) Weinstein DB, Heider JG. Antiatherogenic properties of calcium channel blockers. *Am J Med* 1988; **84**(Suppl 3B): 102–8.

(95) Habib JB, Bossaler C, Wells S, Williams C, Morrisell JD, Henry PD. Preservation of endothelium-dependent vascular relaxation in cholesterol-fed rabbit by treatment with the calcium blocker PN 200–110. *Circ Res* 1986; **58**: 305–9.

(96) Hügenholz PG, Lichtlen P. On a possible role for calcium antagonists in atherosclerosis: a personal view. *Eur Heart J* 1986; **7**: 546–59.

(97) Water D, Freedman D, Lesperance J *et al*. Design features of a controlled clinical trial to assess the effect of a calcium entry blocker upon the progression of coronary artery disease. *Controlled Clin Trials* 1987; **8**: 216–42.

(98) Greer IA. Platelet function and calcium channel blocking agents. *J Clin Pharmacy Ther* 1987; **12**: 213–22.

(99) Ahn YS, Harrington WJ. Calcium channel blockers in thrombotic disease. *Ad Intern Med* 1987; **32**: 137–54.

Natriuretic effects of calcium antagonists: mode of action, duration and clinical implications

G. Leonetti and A. Zanchetti

Istituto di Clinica Medica Generale e Terapia Medica, Università di Milano and Centro di Fisiologia Clinica e Ipertensione, Ospedale Maggiore, Milano, Italy

Although classical vasodilating drugs such as hydralazine, guancydine and minoxidil are effective in lowering peripheral and renal vascular resistances, their use has been limited because of important side-effects and the development of 'pseudo-tolerance' due to water and sodium retention. For these reasons these classical vasodilators could be employed only in association with diuretics and/or beta-blockers. On the contrary the vasodilating agents of the group of calcium-antagonists are effective and long-lasting antihypertensive agents and, in this presentation, we will review their natriuretic and diuretic effects by considering the mechanism or mechanisms of action, the duration of these effects and finally the potential clinical implications of these metabolic effects.

MECHANISMS OF ACTION

By reviewing the natriuretic and diuretic effects of calcium-antagonists it is correct to start by reporting the pioneeristic results of Klutsch et al. (1) who administered intravenously 1 mg of nifedipine to 9 patients with essential hypertension and to 11 patients with renal hypertension. They found that in patients with essential hypertension there was a brisk and significant rise in urine volume and sodium excretion. Since then numerous studies have confirmed that the acute administration of calcium antagonists in hypertensive patients causes a rise in sodium and water excretion. However, according to our experience (2), the dihydropyridine derivatives are more powerful than the papaverine like calcium antagonists: indeed the rise in sodium and water excretion was significantly increased after the acute administration of nifedipine, whereas the rise in sodium and water after verapamil administration was not statistically significant, although the blood pressure was equally reduced by both calcium antagonists (Fig. 1).

Among the possible mechanisms of the acute natriuretic and diuretic effects the haemodynamic changes at the renal level will be first considered. While renal plasma flow has been frequently, but not invariably found to be increased after acute calcium antagonist administration, glomerular filtration rate has been

The use of isradipine and other calcium antagonists in cardiovascular diseases, edited by P. A. van Zwieten, 1989; Royal Society of Medicine Services International Congress and Symposium Series No. 157, published by Royal Society of Medicine Services Limited.

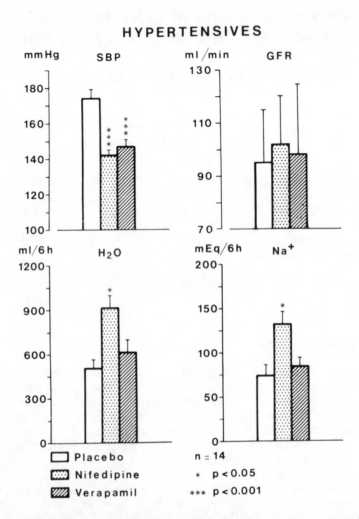

Figure 1 *The effects of acute oral administration of nifedipine 10 mg and verapamil 160 mg on systolic blood pressure (SBP), glomerular filtration rate (GFR), urinary volume (H_2O) and urinary excretion (Na^+) in 14 hypertensive patients (from Leonetti* et al. *with permission) (2).*

reported to be generally unchanged (3–5). In the study of Klutsch (1) it had been shown that the significant rise in sodium and water excretion after intravenous administration of nifedipine was more persistent and independent from the transient rise in glomerular filtration rate (present only in the first 10-minute period). Our own experience (2) with a single oral dose of nifedipine to hypertensive subjects showed a doubling of sodium and water excretion in spite of a non-significant change in glomerular filtration rate from 95 ± 21 to 102 ± 18 ml/minute. Furthermore, animal studies by Dietz *et al.* (6), recently confirmed with the new calcium antagonist amlodipine (7), have shown that intrarenal infusions of nifedipine and verapamil significantly increased sodium and water excretion not associated with significant changes in glomerular filtration rate, renal plasma flow and systemic blood pressure. It is, therefore, unlikely that the diuretic and natriuretic effect of calcium antagonists could be due to an increase in filtered load and it also seems unlikely that an increase in renal blood flow, which occurs with sodium retaining vasodilators as well, plays any significant role. On the contrary, when too large doses of calcium antagonists are administered, the reduction in glomerular filtration rate can overcome the diuretic and natriuretic effects of calcium antagonists in spite of no further blood pressure reduction: indeed (8) when high doses of felodipine as 12.5, 25 or 50 mg t.i.d. were administered, there was a slight reduction in glomerular filtration rate (which was statistically significant with the 50 mg t.i.d. dose only), together with a concomitant reduction in sodium excretion.

Interference with renin release might be a mechanism through which calcium antagonists exert their natriuretic action. However, a decrease in plasma renin activity is rarely reported with verapamil or its derivative methoxyverapamil, the more frequent response being no change at all. Furthermore, acute administration of nifedipine, which causes a greater increment in renal excretion than verapamil, increases rather than decreases plasma renin activity (9,10). Similar results, that is no change in renin release despite a marked increase in urine flow and sodium excretion were found in animal experiments during intrarenal verapamil infusion (6). Interference with renin secretion is, therefore, an unlikely mechanism of the natriuretic response to calcium antagonists.

It is equally unlikely that the increment in sodium excretion could be due to a suppression in aldosterone secretion: acute administration of both nifedipine and verapamil in humans has not been found to be associated with a decrease in plasma or urinary aldosterone. Furthermore, repeated administration of nifedipine does not, according to our personal experience (11), negatively interfere with the adrenal gland secretion of aldosterone in response to physiologic stimuli as angiotensin II and/or potassium infusions, although the rise in plasma renin activity during acute administration of dihydropyridine calcium antagonists is not paralleled by a proportional increase in plasma aldosterone.

Calcium antagonists could cause their natriuretic effect by interfering with the tubular sodium reabsorption mediated by the sympathetic nervous system (12–14). Indeed there is increasing evidence that the sympathetic nervous system, through alpha-receptors, can enhance tubular sodium retention and on the other hand calcium antagonists have been shown to block alpha-receptor mediated calcium channels (15). In spite of these physiologic premises it is unlikely that this mechanism may account entirely for the natriuretic action of calcium antagonists; indeed it has recently been shown that diltiazem and nifedipine do not impair the tubular sodium reabsorption of the rat kidney during modest renal nerve stimulation (16) and that both compounds can exert an acute diuretic and natriuretic action on the denervated rat kidney (17). Recently Johns (7) has shown that a new calcium antagonist, amlodipine, despite causing a significant diuretic and natriuretic effect, does not interfere with tubular sodium reabsorption following a low-frequency renal nerve stimulation.

Rappelli et al. (18) have also shown that nifedipine induced diuretic and natriuretic effect is not mediated by the activation of the kinin-kallikrein system.

Finally, the possibility that calcium antagonists directly inhibit sodium reabsorption has to be carefully considered: indeed there is substantial evidence in that direction, although it is mostly based on animal experiments and some caution must be kept in transferring these results to clinical situations. According to micropuncture studies (19) the effect of calcium antagonists on sodium reabsorption should be at the level of distal tubule, both early and late distal tubular sites, although it has not been excluded that, according to other authors (20), there is a possible action at the level of the proximal tubule.

DURATION OF NATRIURETIC ACTION

While the natriuretic and diuretic effects of the acute administration of calcium antagonists have been investigated in many studies and practically with all agents belonging to this class of drugs, the duration of these effects has been evaluated only in a few studies, generally of relatively short duration due to the difficulty in keeping the patients on a constant daily sodium intake. According to our experience (21) the repeated verapamil administration for 10 days did not show any late reduction in water and sodium excretion: indeed both during the early (from day 1 to day 5) and late (from day 6 to day 10) phases of the study the sodium and water excretion were slightly greater than during placebo, while body weight was, if any, slightly but significantly reduced (Fig. 2). Our results agree with those of Pessina's group (22), who showed a significant decrease in body weight and interstitial fluid volume after 4 weeks of verapamil treatment.

In a sodium balance study with repeated felodipine administration for a 7-day period (23) we calculated a cumulative negative sodium balance of about 80 mEq during the 1st and 2nd day of felodipine treatment and although no further significant natriuretic effect was observed from day 3 onward no rebound retention was observed either, so that the negative sodium balance was maintained until the last (7th) day of felodipine administration. Luft et al. (24) during repeated nitrendipine administration found a similar cumulative negative sodium balance (-103 ± 25 mEq in 8 days), although the pattern of the cumulative balance was slightly different from that observed in our study, probably as a result of the schedule of drug administration.

In order to investigate the long-term effects of nifedipine in sodium balance MacGregor et al. (25) performed a balance study of sodium after withdrawal of nifedipine administered for a period ranging from 6 to 45 weeks. They found during the 7-day period following nifedipine withdrawal a positive balance of 137 ± 14 mEq together with an increase of 0.7 kg of body weight. When nifedipine administration was resumed there was a subsequent loss of sodium and reduction in body weight.

Altogether the various studies suggest that calcium antagonists cause an acute loss of sodium which is maintained in the long-term treatment in a similar fashion of the sodium loss seen during treatment with thiazide diuretics.

CLINICAL IMPLICATION OF NATRIURETIC ACTIVITY

While the antihypertensive effect of calcium antagonists is undoubtedly due to a reduction in total peripheral vascular resistance, the role of the diuretic and natriuretic effects cannot be judged in detail. We have investigated (26) this problem during the acute administration of nifedipine in hypertensive patients kept on normal and low sodium diets respectively. Both during sodium repletion and depletion nifedipine significantly reduced blood pressure. However, although the absolute blood pressure reduction after nifedipine was greater during

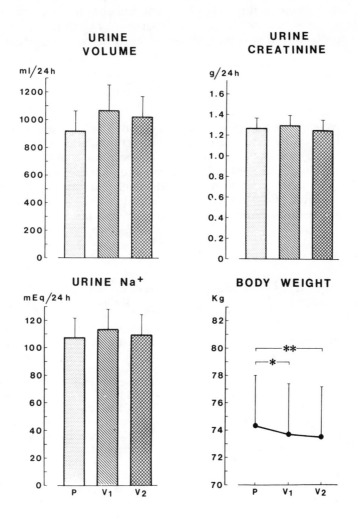

Figure 2 *Effects of verapamil on urine volume (ml/24 hour), urine creatinine (g/24 hour), sodium excretion (mEq/24 hour) and body weight (kg) during the placebo period (P) and two consecutive periods of verapamil administration (V1 and V2). Means ± SE of 5-day values in 12 patients during each period. *p<0.05, **p<0.01 (from Leonetti et al., with permission) (2).*

sodium repletion than during sodium depletion, it seems of interest to emphasize two aspects:

1. the blood pressure values after nifedipine administration were slightly lower during sodium depletion than during sodium repletion.
2. while nifedipine caused a statistically and clinically significant diuretic and natriuretic effect during sodium repletion, there was no significant rise in urine volume and urinary sodium excretion during sodium depletion.

These results suggest to us that the acute antihypertensive effect of calcium antagonists is not related to the natriuretic and diuretic actions. However, the negative sodium balance in the early phase and the absence of a late retention may explain the lack of development of 'pseudotolerance' frequently found with other vasodilators in particular those with a 'direct' action (hydralazine etc.).

CONCLUSIONS

In conclusion we can say that (i) calcium antagonists have an acute natriuretic and diuretic effect, which is probably due to a direct tubular action of the drugs; (ii) although the duration of these effects has not been adequately investigated, the available data do not suggest a late sodium and water retention; and (iii) it is unlikely that the natriuretic and diuretic effects play an important role in the acute antihypertensive effect of calcium antagonists, while it may prevent the development of pseudotolerance during long-term treatment.

REFERENCES

(1) Klutsch K, Schmidt P, Grosswendt J. The effect of Bay a 1040 on renal function in hypertensive patients (Der Einfluss von Bay a 1040 auf die Nierenfunktion der Hypertonikers). *Arzneim-Forsch* 1972; **21**: 377–80.
(2) Leonetti G, Cuspidi C, Sampieri L, Terzoli L, Zanchetti A. Comparison of cardio-vascular, renal and humoral effects of acute administration of two calcium channel blockers in normotensive and hypertensive subjects. *J Cardiovasc Pharmacol* 1982; **4**: 319–24.
(3) Bauer JH, Sunderrajan S, Reams G. Effects of calcium entry blockers on renin-angiotensin, aldosterone system, renal function and hemodynamics, salt and water excretion and body fluid composition. *Am J Cardiol* 1985; **56**: 62H–7H.
(4) Romero JC, Ray L, Granger JP, Ruilope LM, Rodicio JL. Multiple effects of calcium entry blockers on renal function in hypertension. *Hypertension* 1987; **10**: 140–51.
(5) Loutzenhiser RD, Epstein M. Renal hemodynamic effects of calcium antagonists. *Am J Med* 1987; **82** (Suppl 3B): 23–8.
(6) Dietz JR, Davis JO, Freeman RH, Villareal D, Echtenkamp SF. Effects of intrarenal infusion of calcium entry blockers in anesthetized dogs. *Hypertension* 1983; **5**: 482–8.
(7) Johns EJ. A study of the action of amlodipine on adrenergically regulated sodium handling by the kidney in normotensive and hypertensive rats. *Br J Pharmacol* 1988; **93**: 561–8.
(8) Leonetti G, Gradnik R, Terzoli L, Fruscio M, Rupoli L, Zanchetti A. Felodipine, a new vasodilating agent: blood pressure, cardiac, renal and humoral effects in hypertensive patients. *J Cardiovasc Pharmacol* 1984; **6**: 392–8.
(9) Lederballe Pedersen O, Mikkelsen E, Christensen NJ, Kornerup HJ, Pedersen EB. Effect of nifedipine on plasma renin activity, aldosterone and catecholamines in arterial hypertension. *Eur J Clin Pharmacol* 1979; **15**: 235–40.
(10) Schoen RE, Frishman WH, Shammoon H. Hormonal and metabolic effects of calcium channel antagonists in man. *Am J Med* 1988; **84**: 492–504.
(11) Terzoli L, Leonetti G, Pedretti R *et al*. Nifedipine does not blunt the aldosterone and cardiovascular response to angiotensin II and potassium infusion in hypertensive patients. *J Cardiovasc Pharmacol* 1988; **11**: 317–20.
(12) Bello-Reus E, Trevino DL, Gottschalk CW. Effect of sympathetic renal nerve stimulation on proximal water and sodium reabsorption. *J Clin Invest* 1976; **57**: 1104–7.
(13) Di Bona GF. The function of renal nerves. *Rev Physiol Biochem Pharmacol* 1982; **94**: 75–181.

(14) Osborn A, Holdaas H, Thomas MD, Di Bona GF. Renal adrenoceptor mediation of antinatriuretic and renin secretion responses to low frequency renal nerve stimulation in the dog. *Circ Res* 1983; **53**: 298–305.

(15) Van Zwieten PA, Van Meel JCA, Timmermans PBMWM. Pharmacology of calcium entry blockers: interaction with vascular alpha-adrenoceptors. *Hypertension* 1983; **5** (Suppl II): 8–17.

(16) Johns EJ. The influence of diltiazem and nifedipine on the response of the rat kidney to modest renal nerve stimulation. *Br J Pharmacol* 1984; **82**: 261–6.

(17) Johns EJ. The effect of acute administration of diltiazem and nifedipine on the function of denervated kidney. *Br J Pharmacol* 1984; **82**: 328–33.

(18) Madeddu P, Oppes A, Soro A *et al*. Natriuretic effect of acute nifedipine administration is not mediated by the renal kallikrein-kinin system. *J Cardiovasc Pharmacol* 1987; **9**: 536–40.

(19) Di Bona GF, Sawin LL. Renal tubular site of action of felodipine. *J Pharmacol Exp Ther* 1984; **228**: 420–9.

(20) Abe Y, Komori T, Miura K. Effect of the calcium antagonist nicardipine on renal function and renin release in dogs. *J Cardiovasc Pharmacol* 1983; **5**: 254–9.

(21) Leonetti G, Sala C, Bianchini C, Zanchetti A. Antihypertensive and renal effects of orally administered verapamil. *Eur J Clin Pharmacol* 1980; **18**: 375–82.

(22) Semplicini A, Pessina AC, Rossi GP *et al*. Plasma levels of verapamil and its effects on blood pressure, body fluid volumes and renal function in hypertensive patients. *Internat J Clin Pharm Res* 1982; **2** (Suppl 1): 81–6.

(23) Leonetti G, Gradnik R, Terzoli L *et al*. Renal and antihypertensive effects of felodipine in hypertensive patients. *J Hypertension* 1985; **3** (Suppl 3): S161–3.

(24) Luft FC, Aronoff GR, Sloan RS, Fineberg NS, Weinberg MH. Calcium channel blockade with nitrendipine: effects on sodium homeostasis, the renin-angiotensin system, and sympathetic nervous system in human. *Hypertension* 1985; **7**: 438–42.

(25) MacGregor G, Pevahouse JB, Cappuccio FP, Markandu MD. Nifedipine, diuretics and sodium balance. *J Hypertension* 1987; **5** (Suppl 4): S127–31.

(26) Leonetti G, Rupoli L, Gradnik R, Zanchetti A. Effects of a low-sodium diet on antihypertensive and natriuretic responses to acute administration of nifedipine. *J Hypertension* 1987; **5** (Suppl 4): S57–60.

Low dose isradipine MR once daily controls diurnal blood pressure in essential hypertension

W. Diemont, J. Beekman, A. M. J. Siegers and J. H. B. de Bruijn

Department of Internal Medicine, Medisch Spectrum Twente,
Ariënsplein 1, 7511 JX Enschede, The Netherlands.

SUMMARY

In twenty-two patients with mild to moderate essential hypertension the effects of isradipine MR once daily on diurnal blood pressure and heart rate were studied, using a non-invasive ambulatory blood pressure monitor (Spacelabs 5200).

Both isradipine MR 5 mg and 10 mg once daily were shown to reduce diurnal blood pressure throughout the 24-hour observation period ($p < 0.001$). The magnitude of the average diurnal reduction of mean blood pressure was greater on isradipine MR 10 mg (11 ± 2 mmHg) than on isradipine MR 5 mg (8 ± 1 mmHg $p < 0.01$). However, no difference in the duration of the effect could be demonstrated between the two doses. Isradipine MR did not cause reflex tachycardia. Side-effects of the drug were mild and never led to a discontinuation of the drug. Low dose isradipine MR once daily effectively controls 24-hour blood pressure in patients with essential hypertension.

INTRODUCTION

Calcium antagonists are now widely used as antihypertensive agents. In previous studies we and others have demonstrated the safety and efficacy of isradipine (a new dihydropyridine calcium antagonist) (1,2). In a recent study we compared the relative efficacy of identical daily doses of isradipine twice daily and isradipine MR once daily in a small number of patients. Both forms of the drug were shown to lower 24-hour blood pressure effectively, but the 24-hour profile of blood pressure reduction by isradipine MR seemed to have some advantages over isradipine twice daily (3).

However, in the light of our present knowledge of the side-effects/efficacy profile of isradipine, the doses we used in that study were rather high (1). In the present study we set out to investigate the efficacy and safety of low doses (5 and 10 mg once daily) of isradipine MR on diurnal blood pressure compared with placebo.

The use of isradipine and other calcium antagonists in cardiovascular diseases, edited by P. A. van Zwieten, 1989; Royal Society of Medicine Services International Congress and Symposium Series No. 157, published by Royal Society of Medicine Services Limited.

PATIENTS AND METHODS

Twenty-four patients (age 31–73 years, 13 female) with mild to moderate essential hypertension (sitting diastolic office blood pressure on placebo > 100 mmHg) were studied after they had given their informed consent to the study protocol and procedures.

After a single blind placebo run in period of 3 weeks, the patients entered the double blind phase and were randomly assigned to either placebo or isradipine MR 5 mg, or isradipine MR 10 mg once daily, which was given for 4 weeks. The patients who received active treatment during these 4 weeks were subsequently treated with the alternate dose (5 or 10 mg) for 4 additional weeks. The patients who had been treated with placebo during this 4 week period were then randomly assigned to either 5 mg or 10 mg isradipine MR for 4 additional weeks and finally after this period were given the alternate dose of isradipine during a further 4 week period. Accordingly, 8 patients were treated with placebo for two periods, 24 patients were treated with placebo and both doses of isradipine (Fig. 1). At 2 weekly intervals office blood pressure was measured in the morning, between 8.00 h and 9.00 h, 24 hours after the last oral dose of either isradipine MR or placebo, in the supine, sitting and standing positions and heart rate was recorded. All measurements were made after at least 10 minutes of rest.

At the end of the placebo period and after each treatment period non-invasive 24-hour ambulatory blood pressure was measured using a Spacelabs (model 5200) ambulatory blood pressure monitor as described previously (4). At this time, 24 hours after the last oral dose, samples were taken for the determination of standard haematological values, renal and liver function tests, and plasma concentrations of isradipine.

Two patients could not be analysed properly because of inadequate 24-hour blood pressure readings (one on placebo and one who received isradipine 10 mg initially).

Side-effects in the patients were mild and no reason for drop out. So final analysis was done in 7 patients who received placebo twice (age 47±3 years, 1

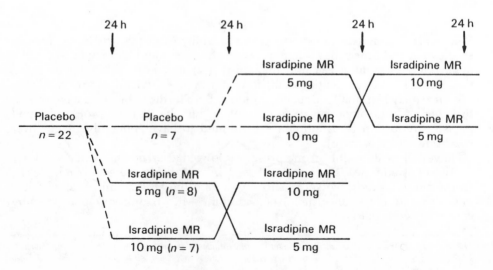

Figure 1 *Study protocol.*

female) and 22 patients on both isradipine MR 5 mg and 10 mg as compared with placebo (51±2 years, 12 female).

Data are given as means±S.E.M. For comparison of results student T-test for paired data was used.

RESULTS

Office blood pressure

In the 7 patients who received placebo twice, sitting office blood pressure at the end of the single blind period was 166±9 mmHg systolic and 105±2 mmHg diastolic and did not change after 4 more weeks placebo (165±9 mmHg/ 101±3 mmHg). No change in heart rate was observed.

In the 22 patients who received both isradipine MR 5 mg and 10 mg once daily office blood pressure decreased from 171±4 mmHg systolic and 107±1 mmHg diastolic on placebo to 157±3 mmHg systolic and 95±2 mmHg diastolic on isradipine MR 5 mg once daily ($n=22$, $p<0.001$) and to 152±4 mmHg and 92±1 mmHg on isradipine MR 10 mg once daily ($n=22$, $p<0.001$).

The difference between isradipine MR 5 mg once daily and isradipine MR 10 mg once daily was also significant ($n=22$, $p<0.05$). Heart rate, on placebo, on isradipine MR 5 mg, and on isradipine MR 10 mg once daily, respectively, appeared to be the same.

Ambulatory blood pressure measurements

In the 7 patients who received placebo twice, average 24-hour ambulatory blood pressure did not change: 147±5/100±4 mmHg vs 144±5/98±3 mmHg. The same holds true when we analysed daytime (8.00–20.00 h) blood pressure only: 154±4/106±3 mmHg vs 149±4/103±3 mmHg. Heart rate was not different either: 80±2 vs 79±3 beats/minute.

Figure 2 *Effects of isradipine MR 5 mg ($\Diamond$, $\triangle$) once daily on 24-hour ambulatory blood pressure compared with placebo ($\Box$, +) in 22 patients with essential hypertension.*

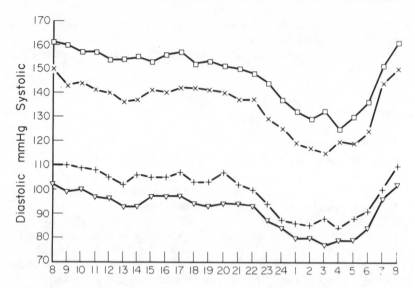

Figure 3   *Effects of isradipine MR 10 mg (×, ▽) once daily on 24-hour ambulatory blood pressure compared with placebo ((□, +) in 22 patients with essential hypertension.*

In the group of 22 patients, average 24-hour ambulatory blood pressure decreased from $147 \pm 3/99 \pm 2$ mmHg on placebo to $137 \pm 3/93 \pm 2$ mmHg on isradipine MR 5 mg ($p < 0.001$) and to $134 \pm 2/91 \pm 2$ mmHg on isradipine MR 10 mg ($p < 0.001$) and average daytime blood pressure (8.00–20.00 h) decreased from $156 \pm 3/106 \pm 2$ mmHg to $144 \pm 3/99 \pm 2$ mmHg on isradipine MR 5 mg ($p < 0.001$) and to $141 \pm 3/97 \pm 2$ mmHg on isradipine MR 10 mg ($p < 0.001$) (Fig. 2, Fig. 3).

The differences between isradipine MR 5 mg and 10 mg once daily are small but statistically significant ($p < 0.01$).

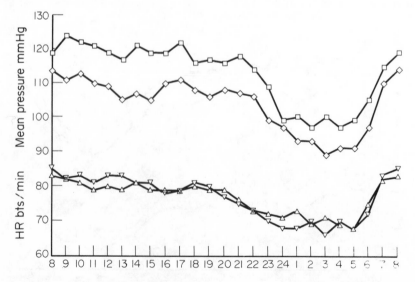

Figure 4   *Effects of isradipine MR 10 mg once daily on 24-hour ambulatory mean arterial pressure (◇) and heart rate (▽) compared with placebo (□, △) in 22 patients with essential hypertension.*

Average 24-hour heart rate was not different between either placebo, isradipine MR 5 mg or isradipine MR 10 mg (77 ± 2 vs 75 ± 1 vs 76 ± 1 beats/minute) (Fig. 4).

DISCUSSION

In the present study both isradipine MR 5 mg and 10 mg are shown to reduce sitting office and 24-hour ambulatory blood pressures in patients with essential hypertension. Given once daily no escape from the effect of the drug can be observed within the 24 hours after the last oral dose, which implies that isradipine given in this modified release preparation is as effective as twice daily dosing of the original form as we have demonstrated previously in a small number of patients (3). The doses used in the present study, however, are lower than those in our previous studies.

With these lower doses side-effects were few and never a reason for discontinuation of the drug. Also the (small) reflex tachycardia observed with the initial higher doses of isradipine (2) is not present any more.

Concomitantly, low dose isradipine MR once daily seems to be an effective and safe means to reduce diurnal blood pressure in patients with essential hypertension.

REFERENCES

(1) Kirkendall WM. Comparative assessment of first line agents for treatment of hypertension. *Am J Med* 1988; **84** (Suppl 3B); 32–41.
(2) Vermeulen A, Wester A, Willemse P, Lustermans FAT, Stegeman CA, de Bruijn JHB. Comparison of isradipine and diltiazem in the treatment of essential hypertension. *Am J Med* 1988; **84** (Suppl 3B): 42–45.
(3) Stegeman CA, Vermeulen A, Wester A, Willemse PA, Lustermans FAT, de Bruijn JHB. Subgroups of calcium antagonists: comparative effects of isradipine and diltiazem in the management of essential hypertension. In: van Zweiten PA, ed. *Calcium antagonists—investigations with isradipine*. London: Royal Society of Medicine Services, 1988: 11–19.
(4) Meyer JL, Ardesch HG, van Rooyen JC, de Bruijn JHB. Low dose captopril twice daily lowers blood pressure without disturbance of the normal circadian rhythm. *Postgrad Med J* 1986; **62** (Suppl 1): 101–5.

Isradipine in ischaemic heart disease and heart failure

E. W. van den Toren, K. I. Lie and W. H. van Gilst

Department of Cardiology, Thoraxcenter, Groningen, The Netherlands

Isradipine is a new dihydropyridine derivate with powerful peripheral and coronary vasodilating properties. In comparison to nifedipine and diltiazem the negative inotropic effects are negligible at equipotent hypotensive dosages, both in animal and clinical studies (1,2). In vitro studies (3) also reveal negative chronotropic effects. This combination of actions may be useful in the treatment of angina pectoris and heart failure. This chapter provides a brief review on the potential value of isradipine in various subsets of coronary heart disease and heart failure.

STABLE ANGINA

Calcium antagonists have been extensively studied in stable exertional angina (4). In these studies verapamil, nifedipine and diltiazem were all effective in reducing spontaneous anginal attacks and exercise induced ischaemia. Rate pressure product was usually slightly reduced, suggesting that improvement of coronary flow to ischaemic zones or reduction of ischaemia by improving intracellular calcium homeostasis may be important antianginal mechanisms of calcium antagonists (4).

The effectiveness of isradipine as monotherapy in stable angina was compared with nifedipine in two randomized double blind cross-over studies (5,6). In 11 patients isradipine 2.5 mg three times daily was contrasted to nifedipine 20 mg three times daily with dose titration at 6-week intervals (5). In the other study 7.5 mg isradipine three times daily was compared with nifedipine 30 mg three times daily (6) in 29 patients.

In both studies isradipine was equipotent to nifedipine in reducing anginal attacks, nitroglycerine consumption and in increasing exercise tolerance when compared with baseline. Haemodynamic changes following active therapy were similar in both groups at higher dosages (6) but showed more pronounced decrease of rate pressure product in isradipine treated patients in comparison with nifedipine at lower dosages (5). When given in higher dosages, isradipine was better tolerated than nifedipine (6).

In both studies no differences occurred in heart rate, suggesting that the potential negative chronotropic effects of isradipine are probably offset by the powerful peripheral vasodilating effects.

The use of isradipine and other calcium antagonists in cardiovascular diseases, edited by P. A. van Zwieten, 1989; Royal Society of Medicine Services International Congress and Symposium Series No. 157, published by Royal Society of Medicine Services Limited.

Another study assessing the role of isradipine when added to patients with beta-blocker therapy is still ongoing (7), but preliminary results in 12 patients reveal an increased exercise time and decreased nitroglycerine consumption in favour of isradipine compared with placebo.

The special features of isradipine observed in preclinical studies are not apparent in these studies; however, it should be realized that most patients in the study mentioned above had normal left ventricular function.

The less outspoken negative inotropic effects of isradipine might be clinically relevant in subjects with angina and compromised left ventricular function. In addition, isradipine can be useful in patients with angina who develop heart failure during treatment with beta-blockers. Further research in these areas is necessary in order to evaluate the potential clinical advantages of isradipine above the conventional calcium entry blockers.

ACUTE ISCHAEMIC SYNDROMES

The rationale for using calcium entry blockers in acute ischaemic syndromes include their anti-ischaemic properties, their possible role in preventing calcium overload, and their ability to prevent coronary spasm and constriction.

Several intervention studies have been carried out in unstable angina and acute myocardial infarction using nifedipine, verapamil or diltiazem (4,8,29). In all studies with either nifedipine, diltiazem or verapamil, no overall reduction of morbidity or mortality was observed in favour of the active drug (4). Only diltiazem used in a specific subset of acute ischaemia or so called non-Q wave infarction was associated with a lower incidence of non-fatal recurrent infarction (8). Furthermore, diltiazem appeared to be beneficial in the subset of patients without pulmonary congestion after AMI which is about 80% of the total infarct population (9). Since the exact mechanism of inefficacy of nifedipine in these acute ischaemic syndromes is not established, it is not known at present whether this inefficacy applies for all dihydropyridines or whether isradipine has additional effects in unstable angina or acute myocardial infarction.

Conversely, all types of calcium entry blockers are effective in variant angina (10). Reduction of complaints have been reported to occur in 80% of cases (10). Although isradipine has not yet been studied in variant angina, there is no reason to assume its inefficacy in view of its powerful vasodilating properties on coronary vascular tone.

CONGESTIVE HEART FAILURE

Vasodilator therapy has become an established approach in the treatment of heart failure (11,12). Since coronary artery disease is the predominant cause of heart failure, isradipine might be particularly useful in view of its powerful peripheral vasodilating actions with concomitant effects on the coronary arteries. In addition, the negative inotropic effects of isradipine are negligible when compared with nifedipine in equipotent antihypertensive dosages.

In heart failure isradipine showed improvement of central haemodynamic parameters in acute and short-term studies (12,13). When administered intravenously to patients with heart failure, isradipine resulted in a decrease of systemic vascular resistance of 29%, an increased cardiac index of 25%, a decrease of pulmonary capillary wedge pressures of 24% without associated changes in heart rate (12). Oral administration of 5 mg and 15 mg isradipine resulted in similar effects on central haemodynamics (13). These findings are in agreement with other acute studies with dihydropyridine calcium antagonists (14). A medium-term (12 weeks) placebo controlled study is still ongoing but long-term controlled studies have not (yet) been performed.

Medium term studies with other dihydropyridine calcium antagonists (15,16) reveal sustained improvement of central haemodynamics, but only marginal improvement of exercise performance and VO_2 max.

Factors such as lack of effect on venous tone, or intrinsic negative inotropic effects might explain these findings. Still, in the absence of other medium- or long-term data, no definite conclusion can be drawn concerning the long-term efficacy of calcium antagonists in patients with congestive heart failure.

ARRHYTHMIAS

Calcium antagonists, other than dihydropyridines, can control supraventricular arrhythmias due to slowing conduction through the atrioventricular node (17).

In this regard isradipine behaves as a true dihydropyridine, and does not influence sinus node activity nor AV conduction, when given either in combination with a beta-blocker, or alone (18).

Therefore, no additional antiarrhythmic effects, other than secondary to the anti-ischaemic mechanism, may be expected from isradipine.

ANTIATHEROGENIC PROPERTIES

Remarkable consistency has been seen in the effect of all classes of calcium antagonists to slow atherogenesis in cholesterol fed animals (19). Isradipine was shown to slow atherogenesis in rabbits (20). Nevertheless, no data on antiatherogenic properties in man of any of the calcium antagonists are available, and further studies should be undertaken.

CONCLUSION

It may be concluded that isradipine is a useful drug in the treatment of stable angina as monotherapy or in combination with beta-blockers. Long-term studies in heart failure are necessary to assess its efficacy in heart failure, but no aggravation of existing heart failure can be expected when adding isradipine to standard therapy.

Isradipine has no place in the treatment of supraventricular arrhythmias, since no decrease in AV conduction has been shown, neither alone, nor in combination with a beta-blocker. Its effect on the slowing of atherosclerosis in man has yet to be elucidated.

REFERENCES

(1) Hof RP. Comparison of cardiodepressant and vasodilator effects of isradipine, diltiazem and nifedipine in anesthetized rabbits. *Am J Cardiol* 1987; **59**: 37B–42B.

(2) Mauser M, Voelker W, Karsch KR. Isradipine, ein neuer Dihydropyridine Kalzium-antagonist mit geringeren negativ inotropen Eigenschaften im Vergleich zu Nifedipin. *Z Kardiol* 1988; **77**: 373-7.

(3) Hof RP, Scholtysik G, Loutzenhiser R, Vuoerela HJ, Neumann P. PN 200-110, a new calcium antagonist; electrophysiological inotropic and chronotropic effects on guinea pig myocardial tissue and effects on contraction and calcium uptake of rabbit aorta. *J Cardiovasc Pharmacol* 1984; **6**: 399-406.

(4) Lie KI. Calcium entry blockers and ischemic heart disease. In: Van Zwieten PA, ed. *Calcium antagonists—investigations with isradipine.* London: Royal Society of Medicine, 1988: 39-42.

(5) Handler CE, Rosenthal E, Tsagadopoulos D, Najm Y. Comparison of isradipine and nifedipine in chronic stable angina. *Int J Cardiol* 1988; **18**: 15-26.

(6) Pool PE, Seagren SC, Salel AF. Isradipine in the treatment of angina pectoris. *Am J Med* 1988; **84** (Suppl 3B): 62-6.

(7) Van den Toren EW, van Bruggen A, Ruffman K, Lie KI. Safety and efficacy of isradipine in angina pectoris patients insufficiently responding to betablocker therapy. In: Van Zwieten PA, ed. *Calcium antagonists—investigations with isradipine.* London: Royal Society of Medicine, 1988: 57-64.

(8) Gibson RS, Boden WE, Theroux P. Dilitiazem and reinfarction in patients with non-Q wave myocardial infarction. Results of a double blind randomized multicenter trial. *New Engl J Med* 1986; **315**: 423-8.

(9) The Multicenter Diltiazem Postinfarction Trial Research Group: The effect of diltiazem on mortality and reinfarction after myocardial infarction. *NEJM* 1988; **319**: 385-92.

(10) Maseri A, Crean PA. Clinical experience with calcium antagonists in angina. In: Opie LH, ed. Calcium antagonists and cardiovascular disease. New York: Raven Press, 1983: 215-20.

(11) Cohn JN, Franciosa J. Vasodilator therapy of cardiac failure. *New Engl J Med* 1977; **297**: 27-31.

(12) van den Toren EW, van Bruggen A, Ruegg P, Lie KI. Hemodynamic effects of iv infusion of isradipine in patients with congestive heart failure. *Am J Med* 1988; **84** (Suppl 3B): 97-101.

(13) Greenberg BH, Siemenczak D, Broudy D. Isradipine improves cardiac function in congestive heart failure. *Am J Med* 1988; **84** (Suppl 3B): 56-61.

(14) van den Toren EW, Lie KI, van Gilstwh. Calcium antagonists and heart failure. *Prog Pharmacol*, in press.

(15) Timmis AD, Smyth P, Kenny JF. Effects of vasodilator treatment with felodipine on haemodynamic responses to tredmill exercise in congestive heart failure. *Br Heart J* 1984; **52**: 314-20.

(16) Dunselman P, Kuntze GE, Lie KI. Efficacy of felodipine in congestive heart failure. *Eur Heart J*, in press.

(17) Zipes DP, Fischer JC. Effects of agents which inhibit the slow calcium channel on sinus node automaticity and AV conduction in the dog. *Circ Research* 1974; **34**: 184-92.

(18) van Wijk LM, van den Toren EW, van Gelder I, Crijns HJGM, Lie KI. Electrophysiological properties of intravenously applied isradipine (PN 200-110) in man with normal sinus node and AV nodal function. Influence of beta-blockade. In: van Zwieten PA, ed. *Calcium antagonists—investigations with isradipine.* London: Royal Society of Medicine, 1988: 75-82.

(19) Winnifred C. Nayler, ed. *Calcium Antagonists.* Academic Press, 1988: 235.

(20) Habib JB, Bossaller C, Hendry PD. Suppression of atherogenesis in cholesterol-fed rabbits with a low dosed calcium antagonist (PN 200-110). *JACC* 1986: **7**: 58A.

Calcium antagonists in the treatment of Raynaud's phenomenon. Discrepancy between short-term and long-term effectiveness

T. Thien and H. Wollersheim

Department of Medicine, Division of General Internal Medicine, St Radboud University Hospital, Geert Grooteplein Zuid 8, 6500 HB Nijmegen, The Netherlands

SUMMARY

The effects of two dihydropyridine type calcium antagonists in the treatment of Raynaud's phenomenon were studied. Both short- and long-term effects, both subjective and objective parameters were established. A single sublingual dose of 10 mg nifedipine in 16 Raynaud patients caused a clear vasodilation, both systemic and in the skin. The recovery after a cold provocation (waterbath of 16 °C, during 5 minutes) was improved against the placebo values. However, after chronic treatment with nifedipine (40 mg daily) for 4 weeks the objective parameters did not show changes, despite some subjective improvements mentioned in the diary. The group was too small to show differences between patients with primary and with secondary Raynaud's phenomenon.

In a second study the effects of an infusion with nicardipine (5 mg/hour) were compared with those of a placebo infusion in 12 Raynaud patients and 12 healthy volunteers. Again the systemic effects reflected a clear vasodilatation. The peripheral skin blood flow parameters showed an improvement. This effect was most pronounced in the group with a primary Raynaud's phenomenon. When given for 3 weeks in an oral dose of 90 mg daily to 24 patients, the objective and subjective effects appeared to be virtually absent. One should take into account that the plasma levels reached were below the therapeutic levels in nearly 50% of the patients.

The side-effects with both calcium antagonists were minor and as to be expected: palpitations, flushing, headache, generally not experienced as severe.

We conclude that the objective improvements exerted by chronic use of calcium antagonists in the treatment of Raynaud's phenomenon are disappointing.

The subjective amelioration was not impressive either. Perhaps there may be some place for these drugs in the prevention of severe attacks, when patients know the stimuli that induce vasospasms.

The use of isradipine and other calcium antagonists in cardiovascular diseases, edited by P. A. van Zwieten, 1989; Royal Society of Medicine Services International Congress and Symposium Series No. 157, published by Royal Society of Medicine Services Limited.

INTRODUCTION

In 1862 Maurice Raynaud described the clinical syndrome, that now bears his name (1). Despite its longstanding description and the improvement of diagnostic procedures, both in measurement of skin blood flow and in the immunological techniques that help to prove or to exclude secondary causes, up to now the pathophysiology still remains unknown and there is no definite cure either. Since the clinical picture is dominated by visible discolorations, due to vasoconstriction or vasospasm, the use of vasodilators has been extensively explored, however, without unequivocal success.

The development of calcium antagonists has offered a new approach. Ultimately the contraction of vascular smooth muscle is dependent on the intracellular calcium concentration, for the greater part determined by the influx of calcium ions through the slow calcium channels. Indeed calcium antagonists have successfully been used in the treatment of vasospastic syndromes like for instance Prinzmetal or variant angina and migraineous headache. Along this line calcium antagonists have also been tried in the treatment of patients with Raynaud's phenomenon (RP). In a number of adequately designed studies, reviewed by Smith and Rodeheffer (2), this treatment was reported to be effective. From these experiences calcium antagonists of the dihydropyridine (nifedipine) type appeared to be most promising.

Since we have collected at our out-patient clinic over 350 patients with rather severe complaints caused by Raynaud's phenomenon, we decided to perform an open pilot study to establish both the acute and chronic effects of nifedipine. Thereafter we studied the effects of another dihydropyridine calcium antagonist namely nicardipine, both after intravenous and chronic oral administration, in a double blind placebo controlled trial. In the intravenous part of the study the effects of nicardipine were also established in a group of healthy volunteers.

PATIENTS

Table 1 shows the clinical characteristics of the patients selected for the studies presented. All patients gave written informed consent and the protocols were

Table 1   *Clinical data of the patients with Raynaud's phenomenon (RP) and of the healthy volunteers (HV): mean ± SD are given*

	Nifedipine study (acute and chronic)	Nicardipine study Acute RP	HV	Chronic
Number	16	12	12	25
Age (years)	46 ± 15	38 ± 13	27 ± 8	41 ± 11
Sex (F/M)	10/6	7/5	7/5	15/10
RP – etiology				
primary	8	6		16
secondary	8[a]	6[b]		9[b]
– duration (years)	8 ± 7	9 ± 9		6 ± 6
– severity (mean daily number of attacks)	6 ± 2	8 ± 5		7 ± 3
– trophic skin lesions	5[c]	4[c]		8[c]

[a]*All had progressive systemic sclerosis (PSS);* [b]*a mixture of connective tissue diseases (no PSS);* [c]*16 out of the total number of 17 patients with skin lesions in the 3 studies belong to the secondary RP group.*

approved by the local Ethical Committee. The patients had to be without other manifest cardiovascular disease and without drugs that could interefere with vascular tone. The patients should have at least one attack per day and since the studies were performed during the winter months actually the lowest number was three attacks daily and the upper number amounted up to 20 attacks. The classification into primary (PRP) and secondary (SRP) Raynaud's phenomenon was done using the criteria of Allen and Brown (3) for PRP, and ARA or other accepted clinical criteria for the different forms of SRP. For immunological examination recently developed and described methods were used (4,5). Before admission the patients had to prove that they were able to keep a standardized diary during a training period of 4 weeks.

METHODS

Subjective measurements

The patients had to record in the diary the daily number of attacks, their mean duration in minutes and the severity on a ten-point scale. At the end of the study the patients chose which treatment period they preferred. Simultaneously side-effects were noted on a preassigned check-list with 40 items of possible and non-existing control adverse reactions. Daily ambient temperature at 4 pm was noted (National Weather Service). In the chronic nicardipine study compliance was determined by asking and by tablet counting.

Objective measurements

At predetermined visits a number of measurements were performed under strictly standardized conditions. All these were conducted in a room with constant temperature (24.8 ± 0.2 SD °C) and humidity (46%), after at least a 20-minute acclimatization period, in the supine position and always after abstaining from smoking for 24 hours (6) and from caffeine- and alcohol-containing beverages for 12 hours. On the left arm blood pressure, systolic and diastolic (SBP and DBP) were measured non-invasively with an Arteriosonde (mmHg). Further forearm blood flow (FBF) was measured with a strain gauge plethysmography (ml/100 ml/minute), while the hand was excluded by a cuff at a pressure of 50 mmHg. Heart rate (HR) was calculated from the ECG in beats per minute (bpm). Mean arterial pressure (MAP) was calculated as the sum of the diastolic and one-third of the pulse pressure and the forearm vascular resistance (FVR) by dividing MAP through FBF, expressed in arbitrary units (AU).

On the distal volar surface of the second, third and fourth finger we respectively measured: finger skin temperature (FST) in °C by a thermocouple, finger skin blood flow by a laser Doppler periflux (LDF) (Perimed) in AU and transcutaneous oxygen tension ($TcPO_2$) with the skin below the electrode heated up to 45 °C, in mmHg by a Tacomette (Novametrix).

The measurements started after the equilibration period. Always a number of baseline readings were taken and averaged before the intervention took place. This consisted of the ingestion of a capsule (nifedipine or placebo) or an infusion (nicardipine or placebo), always followed by a provocation test, by the finger cooling test (FCT), a cold challenge according to Cleophas et al. (6). The gloved right hand was immersed into a waterbath of 16 °C for 5 minutes. Before, during and for 20 minutes after the FCT, all peripheral parameters were measured at regular intervals.

Protocol

In the nifedipine study the effects of the single sublingual dose were measured after a 4-week control period and compared with the effects of a single blind administered placebo capsule. Thereafter, in an open study, the patients used oral nifedipine 10 mg q.i.d. for 4 weeks and at the end again all measurements were repeated. During both 4 week periods the diary had to be kept daily.

In the nicardipine infusion study, the drug was administered in a randomized double blind placebo controlled manner to all 25 subjects (Table 1). The rate of infusion was kept constant at 5 mg/hour for 85 minutes. After 60 minutes infusion the FCT was performed, as described above.

The chronic nicardipine study had a randomized double blind placebo controlled crossover design. Oral nicardipine was used at a dose of 30 mg t.i.d. for 3 weeks. During both 3-week periods the diary was filled in. At the end of the infusion and the chronic drug period blood was withdrawn for determination of the plasma levels of nicardipine (7).

STATISTICAL ANALYSIS

Significance was evaluated by Student's t-test for normally distributed parameters and with Wilcoxon's rank sum test for parameters which are distributed irregularly. Results are expressed as means $\pm$ SE unless indicated otherwise. To obtain a measure for the overall level of FST, LDF and TcPO$_2$ during the FCT the area under the curve from 5 minutes before cooling until the end of the recovery was calculated and divided by time. The resulting quotient is a weighted average of the periods for which each value is a representative. This overall level is called 'mean level during test' (8). p-values below 0.05 (two-sided) were considered significant.

RESULTS

Figure 1 shows the systemic effects of 10 mg nifedipine sublingually, compared with the course of the parameters after placebo. Nifedipine induced the expected vasodilatory effects, with a small but significant decrease of the DBP and a compensatory increase in HR. There was a clear rise in FBF and consequently a fall in FVR.

Figure 2 shows the course of the parameters of the peripheral skin circulation, before and after the sublingual gift of nifedipine, and also during and after the FCT, performed 25 minutes after the intake of the drug. LDF already increased substantially before the FCT. The recovery after cold challenge was also better, although the precooling values were not achieved. The TcPO$_2$ values were higher after the drug than after placebo, before, during and after the FCT. The FST showed only a better recovery after the FCT when the active drug was used. Again, however, the precooling values were not achieved.

Regarding the effects of chronic nifedipine, no significant changes in blood pressure were seen. HR increased significantly in the upright position only, from 80 ± 2 to 85 ± 2 bpm. The effects of chronic nifedipine on the skin circulatory parameters are given in Fig. 3. The 'mean level during test' for FST, LDF and TcPO$_2$ improved significantly after the single dose, but no longer so after the

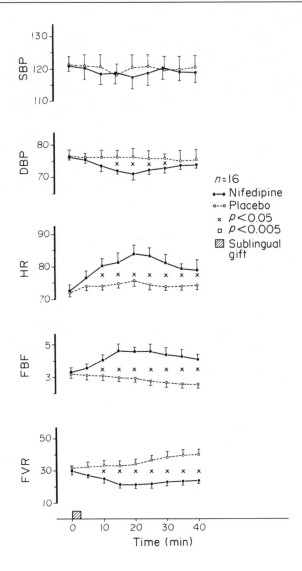

Figure 1 *Effects of a single sublingual dose of nifedipine (10 mg) on haemodynamics on 16 patients with Raynaud's phenomenon. Comparison with placebo. SBP and DBP: systolic and diastolic blood pressure (mmHg); HR: heart rate (beats per minute); FBF: forearm vascular resistance in arbitrary units.*

chronic treatment. Only the mean LDF showed a tendency towards a higher value during nifedipine ($p < 0.07$ against single placebo).

Table 2 shows the results of the diary. There was a significant, although not impressive improvement, with a remaining average of 4.8 attacks per day. The duration of the attacks was not significantly diminished. The severity was subjectively classified as being significantly less. When looking separately to patients with PRP and with SRP (all had progressive systemic sclerosis or scleroderma) the greatest changes were seen in the PRP group. There was no difference in daily ambient temperature during the two study periods.

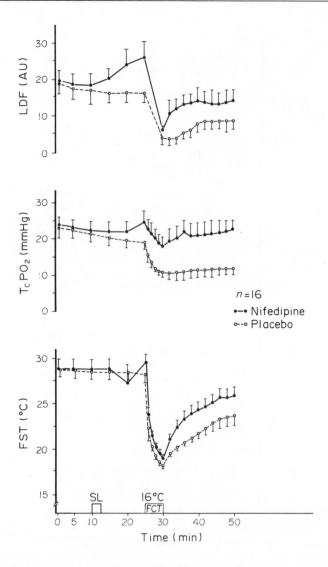

Figure 2 *The peripheral circulatory changes of laser Doppler derived finger skin blood flow (LDF), transcutaneously measured oxygen tension (TcPO₂) and finger skin temperature (FST) before and after 10 mg nifedipine or placebo, both sublingually administered (SL), before and during a fingercooling test (FCT). AU=arbitrary units.*

Figure 4 summarizes the systemic consequences of the intravenous administration of nicardipine. The effects were all as to be expected for an arteriolar vasodilator with a striking reflex rise in HR.

Figure 5 shows the effects of infused nicardipine on FST and LDF in all Raynaud patients. The baseline FST value was lower in the Raynaud group than in the controls, but FST rose significantly in the Raynaud group only. Also the recovery after the FCT improved by nicardipine, almost reaching precooling values. In controls, who already had higher baseline FST values, the changes were small and could only

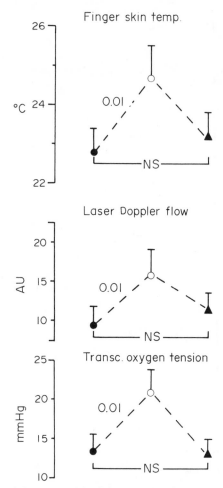

Figure 3 *Effectiveness of chronic use of nifedipine (triangles) against the single sublingual dose of nifedipine (open circles) and against single placebo (closed circles). Given are 'mean levels during test' (8), mean ± SE. AU: arbitrary units; NS: not significant; p-values for differences between single nifedipine and placebo are indicated.*

Table 2 *Diary results in the chronic nifedipine study. Comparison of the eight patients with primary Raynaud's phenomenon with the eight patients with scleroderma (mean ± SEM)*

Attacks	Basic period	p-value	Nifedipine period
Daily number			
Patients with primary Raynaud's phenomenon	4.4±0.5	0.05	3.4±0.6
Patients with scleroderma	7.2±0.8	0.07	6.2±0.6
Duration (minutes)			
Patients with primary Raynaud's phenomenon	26.0±4.3	0.91	26.3±5.2
Patients with scleroderma	35.0±4.9	0.04	29.3±3.5
Severity (0–10 point scale)			
Patients with primary Raynaud's phenomenon	5.8±0.5	0.02	4.6±0.7
Patients with scleroderma	5.4±0.4	0.14	4.6±0.2
Outdoors temperature during 4 weeks (°C)	2.3±0.4	0.96	2.3±0.4

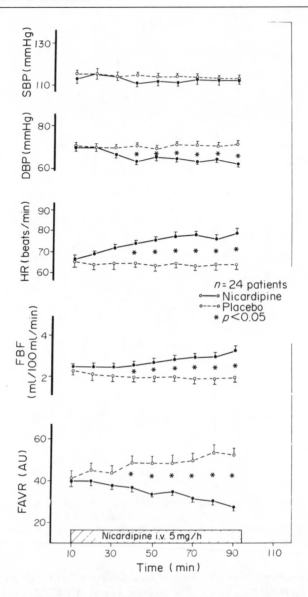

Figure 4 *Effects of nicardipine and placebo infusion on systolic (SBP) and diastolic (DBP) blood pressure, heart rate (HR), forearm blood flow (FBF) and forearm vascular resistance (FAVR), in arbitrary units (AU) in the group of 12 Raynaud patients and 12 healthy volunteers taken together.*

be seen during the recovery phase of the FCT (not shown here). The FST values finally reached were virtually similar in both groups. LDF showed the same tendency as FCT in the Raynaud group but the improvement was less impressive. Particularly the recovery after FCT was clearly better and the precooling values were reached. In the control group the effects were less, but again the baseline values were already higher in comparison with the Raynaud group (40 ± 6 vs 21 ± 9 AU, $p < 0.05$).

Within the group with Raynaud's phenomenon, the beneficial effect of nicardipine appeared mainly to be due to the changes observed in patients with

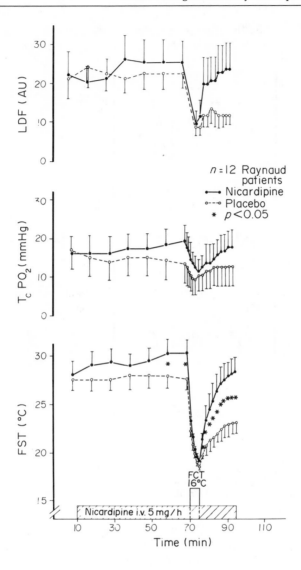

Figure 5 *Effects of nicardipine and placebo infusion of laser Doppler skin blood flow (LDF) in arbitrary units (AU), on transcutaneous oxygen tension (TcPO₂) and on finger skin temperature (FST) in 12 Raynaud patients. At the indicated time point finger cooling test (FCT) was done (for details see methods).*

primary RP. Figure 6 demonstrates that especially the increase of LDF after nicardipine was virtually absent in the subgroup with SRP. Concerning the FST, in the subgroup with SRP there was only an improvement during the recovery phase.

There was no significant change in $TcPO_2$ after nicardipine (Fig. 5). The baseline values were, however, quite different within the groups, at 33 ± 3 in the healthy volunteers, 20 ± 4 in the PRP subgroup and 13 ± 9 mmHg in the SRP subgroup.

After the intravenous study a chronic study was undertaken in which 90 mg nicardipine per day was used during 3 weeks. This study has just been finished

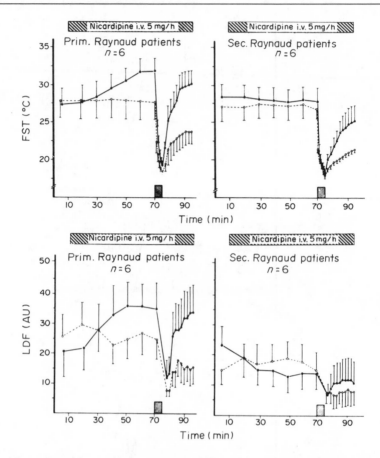

Figure 6 *The course of finger skin temperature (FST) and laser Doppler skin blood flow (LDF) in arbitrary units (AU) during nicardipine (unbroken lines) and placebo infusion (broken lines) in primary (left) and secondary (right) Raynaud patients. At time point 70 min a finger cooling test (FCT), see text, was done.*

and only preliminary results can be presented here. The most notable finding was that in about half of the patients subtherapeutic plasma levels were measured. Figure 7 shows a comparison of the plasma levels found at the end of the oral study and of the infusion study. Plasma levels above 10 ng/ml are considered to have a therapeutic effectivity and this level was not reached in 11 out of 23 patients. The overall effects, both regarding subjective and objective measurements were not significantly different from placebo, but a further analysis with a special emphasis on the drug levels is necessary.

Tablet counting revealed that in 82% (36 out of 44) of the outclinic visits the patients had ingested between 90 and 110% of the prescribed tablets. The mean consumption amounted to 99.8% (range 68–125%). According to the compliance literature this can be interpreted as a rather good success.

SIDE-EFFECTS

Table 3 gives the side-effects noted in both chronic studies. They are according to the known side-effects from calcium antagonists. Again the insufficient

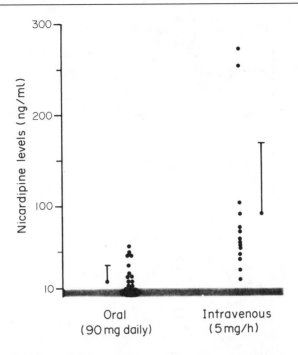

Figure 7 *Plasma levels of nicardipine (ng/ml) achieved at the end of the infusion (right) and at the end of a 3-week oral treatment period (left). The shaded area at the bottom indicates subtherapeutic plasma levels.*

Table 3 *Side-effects of chronic nifedipine and nicardipine*

	Nifedipine 40 mg daily	Nicardipine 90 mg daily	
Total number of patients	16	25	
		Drug	Placebo
Withdrawals due to side-effects	0	2	1
Headache	8	14	10
Flushing	4	8	5
Dizziness on standing	5	5	5
Palpitations	6	6	3
Tiredness	2	4	4
Oedema	2	4	4

plasma concentrations in a substantial number of patients should be taken into account. The side-effects mentioned during placebo affirm the need for a placebo controlled design in clinical drug trials. Despite the fact that some authors have argued that the effects of calcium antagonists on blood pressure and heart rate are virtually absent in normotensives we observed a number of patients with complaints of palpitations. Before the start of the intravenous nicardipine study we applied higher doses in some pilot experiments and had to interrupt the infusion because of sinus tachycardia. Figure 8 shows two examples of volunteers who even had to be treated with a beta-adrenoceptor blocking agent.

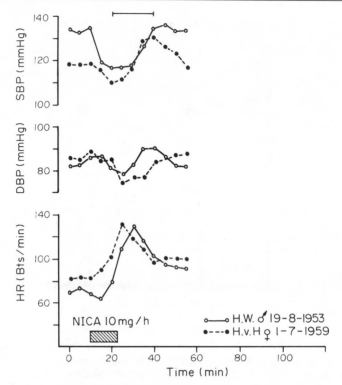

Figure 8 *Pilot study with higher nicardipine (NICA) infusion in two volunteers, who exhibited pronounced side-effects (palpitations, flushing). Effects on systolic (SBP) and diastolic (DBP) blood pressure and on heart rate (HR) in beats per minute (Bts/min).*

DISCUSSION

There is still a need for effective treatment modalities for Raynaud's phenomenon, since all pharmacological regimens are at best palliative and exert a number of side-effects. Any new approach in drug therapy to Raynaud's phenomenon should be tested against placebo and the test should include measurements of both subjective and objective parameters. Then the new drug should be evaluated against other effective therapeutics, thus balancing the effectivity and tolerability of both treatments. In Raynaud's phenomenon there is no unequivocal pharmacotherapy up to now. Table 4 summarizes the possible therapeutic interventions in patients with PRP and SRP. With this background we started to evaluate the application of calcium antagonists for the treatment of RP.

In a pilot study the acute effects of nifedipine were promising, inducing a marked vasodilatation in finger skin and forearm muscle vessels. The results of chronic nifedipine therapy were, however, far less pronounced (9). Moreover, the improvement of the objective parameters after a single dose did not predict the changes after chronic use. In a number of single dose studies in the literature (10,11,12) contradictory results have been reported. Like our nifedipine study not all literature reports had an adequate design. Despite this, the use of a sublingual test with nifedipine as a predictor for chronic treatment was advocated without presenting data (13). In another report both acute and chronic treatment with nifedipine were described (14). Both treatments induced a significant improvement,

Table 4 *Pharmacotherapy of Raynaud's phenomenon*

Vasodilatation
–sympathicolytic drugs: reserpine, alpha-methyldopa
–alpha$_1$-adrenoceptor blockers: phenoxybenzamine, prazosin
–direct acting vasodilators: nitrates, hydralazine,
–serotonin-blockade: ketanserin
–calcium antagonists: nifedipine
–angiotensin converting enzyme inhibitors: captopril
–prostacyclin analogues: iloprost
–thromboxane synthetase inhibitors: dazoxiben

Thrombocyte aggregation inhibitors or haemorrheologic agents
–acetylsalicylate
–buflomedil, isoxsuprine
–pentoxifylline
–dextrane
–stanozolol

but a relationship between acute and chronic effectivity and thus a prediction of a single dose, could not be demonstrated.

Most controlled studies with chronic oral nifedipine show subjective improvement, without a reaction of the objective variables (2). In these studies a relation between subjective alleviation and objective improvement was not found. This may be due to either an insufficient sensitivity of the microcirculatory methods used or to the fact that the cold stimulus is unsuitable as a model for a vasospastic attack. In particular an icewater challenge is an excessive and too painful stimulus in comparison to the mostly mild stimuli experienced during normal daily life.

We always use a mild challenge with a waterbath of 16 °C as provocation. Although the reproducibility is quite acceptable when a standardized protocol is used, the validity as model for a Raynaud attack may be questioned, since we seldom have observed visible colour changes of the fingers with our cold provocation.

The discrepancy between the effects in an acute and a chronic study were also seen in our nicardipine study. In the latter an accidental factor was formed by the too low plasma levels in about half of the patients. This can be explained by an inadequate bioavailability of nicardipine or to a poor compliance. We will have to look further within this study for the possible role of these two factors.

Regarding our methods the FST is thought to represent the total skin blood flow. LDF is also supposed as a parameter for total skin blood flow. It measures the frequency shift of back-scattered laser light by the moving red cells in the outermost layer of the skin (15). LDF correlates well with other methods that determine skin blood flow (16,17,18). The method is non-invasive, easy to repeat, without distress for patients, with low costs and with a sufficient reproducibility. It appears to be a suitable method to follow the effects of acute physiologic and pharmacologic interventions.

TcPO$_2$ (19) is applied as an indicator of capillary perfusion (20), well correlating with other methods (21,22). The nutritional skin blood flow seems particularly affected in patients with secondary forms of the Raynaud's phenomenon, especially in cases with progressive systemic sclerosis (23). Our data confirm the experience in the literature that the beneficial effects of calcium antagonists seem to be largely limited to the PRP patients. The objective peripheral parameters like FST and LDF show the most impressive effects in the groups with PRP. With respect to the TcPO$_2$ the results were contradictory: an improvement after the single dose of nifedipine, whereas hardly any effect could be traced after the nicardipine infusion. During both chronic studies all three parameters of the peripheral circulation remained almost unchanged, with only a marginal amelioration of the LDF in the nifedipine study.

The above mentioned discrepancy between the short- and long-term effectivity has been reported by several authors (2). How to explain this? Firstly, the dose or the bioavailability may have been insufficient. The low plasma levels in the nicardipine support this argument. A second point might be a poor drug compliance of our patients, although we did not find indications for this. The side-effects may have played a role in the potential poor compliance. A third explanation may be that the calcium antagonists cause counter-regulatory mechanisms by activation of the sympathetic and the renin-angiotensin systems. These reactions may stimulate side-effects, but may also induce vasoconstriction. However, the ultimate side-effects noted by the patients were not that impressive and hardly increased against placebo. It is clear that our long-term studies had some limitations. The nifedipine study was open and the dose was fixed. Despite these disadvantages, all the objective measurements did not show significant improvements, whereas open studies generally tend to overestimate the success. Higher daily dosages of nifedipine might have exhibited a greater effectiveness but also an increase of the side-effects (24). In other studies a similar daily dose has proved to be effective (25). The design of the chronic nicardipine study was adequate but again the fixed dose may have been a disadvantage. This is further strengthened by the subtherapeutic levels reached in about 50% of the patients. The literature concerning calcium antagonists in the treatment of Raynaud's phenomenon was reviewed by Smith and Rodeheffer in 1985 (2). From that review and from later studies the opinion emerges that the subjective improvement is greater than the changes in the objective measured parameters. In most chronic studies there was no objectively measured improvement (25,26,27), whereas some reports detected a change in the objective parameters (28,29). From the literature it also appears that the dihydropyridine type calcium antagonists are more effective than the other types, although the other types of calcium antagonists have been dealt with in a rather limited number of reports (2,27,30,31). As already mentioned there is also the tendency that the short-term effectivity is superior to the long-term, with regard to the objective parameters (32). Another striking fact is that the improvement is more impressive in the PRP group (2,27). Both latter points are in accordance with our findings.

On the other hand there are some studies that conclude that calcium antagonists may have beneficial effects in patients with digital ulcerations, caused by scleroderma (33,34). Although we could not measure important changes in the objective parameters in our secondary RP patients, we observed indeed healing of ulcerations in a few severe scleroderma patients treated with nifedipine on an individual basis.

The disappointing results in our chronic nicardipine study are corroborated in one other study using a similar daily dose (35). In that study no significant changes were found neither in the number of attacks per day, nor in the severity of the attacks or other clinical features. Drug levels have not been measured in that study, although a clear suppression of the drug on platelet activation was detected. The latter finding illustrates on one side that the applied dose is able to induce some effects but on the other hand that it seems unlikely that the effects on platelet parameters are directly responsible for the eventual improvement of the Raynaud's phenomenon.

In conclusion the dihydropyridine calcium antagonists exert objective improvement predominantly when applied acutely. In chronic studies this could, however, not be confirmed up to now. In the chronic studies some subjective amelioration was present, without convincing clinical relevance. Finally, only a very low percentage of the patients became completely free of symptoms. Therefore, the treatment of Raynaud's phenomenon remains a clinical challenge. On the other hand the sublingual administration of nifedipine can offer a treatment modality in the prevention of attacks in patients, who are aware of the provoking stimuli for their complaints (36).

REFERENCES

(1) Raynaud, M. *De L'asphyxie locale et de la gangrène symétrique des extrémités*. Paris: Leclerc Rignoux, 1862.
(2) Smith CR, Rodeheffer RJ. Raynaud's phenomenon: pathofysiologic features and treatment with calcium channel blockers. *Am J Cardiol* 1985; **55**: 154B–7B.
(3) Allen E, Brown G. Raynaud's disease. A critical review of minimal requisites for diagnosis. *Am J Med Sci* 1932; **183**: 187–200.
(4) Kallenberg CGM, Wouda AA, Hoet MH, van Venrooy WJ. Development of connective tissue disease in patients presenting with Raynaud's phenomenon. *Ann Rheum Dis* 1988; **47**: 634–41.
(5) Wollersheim H, Thien T, Hoet MH, van Venrooy WJ. The diagnostic value of several immunological tests for ANA in predicting the development of connective tissue disease in patients presenting with Raynaud's phenomenon. (Submitted for publication).
(6) Cleophas T, Fennis J, van 't Laar A. Finger skin temperature after a finger cooling test. *J Appl Physiol* 1982; **52**: 1167–71.
(7) Higushi S, Sasaki H, Sado T. Determination of a new cerebral vasodilator 2,6-dimethyl-4-(3-nitrofenyl)-1,4-dihydropyridine-3, 5 dicarboxylic acid 3-(2-N-benzyl-methylamino)-ethyl-ester-5-methylester-hydrochloride (yc-93) in plasma by electron capture gas chromatography. *J Chromatography* 1975; **110**: 301–7.
(8) Wollersheim H, Thien T, Fennis J et al. Double blind, placebo-controlled study of prazosin in Raynaud's phenomenon. *Clin Pharmacol Ther* 1986; **40**: 219–25.
(9) Wollersheim H, Thien T, van 't Laar A. Nifedipine in primary Raynaud's phenomenon and in scleroderma: oral versus sublingual hemodynamic effects. *J Clin Pharmacol* 1987; **27**: 907–13.
(10) Kahan A, Weber S, Amer B et al. Calcium entry blocking agents in digital vasospasm (Raynaud's phenomenon). *Eur Heart J* 1984; **4** (Suppl C): 123–9.
(11) Winston EL, Pariser KM, Miller KB et al. Nifedipine as a therapeutic modality for Raynaud's phenomenon. *Arthritis Rheum* 1983; **26**: 1177–80.
(12) Wise RA, Malamet R, Wigley FM. Acute effects of nifedipine on digital blood flow in human subjects with Raynaud's phenomenon: a double blind placebo controlled trial. *J Rheumatol* 1987; **14**: 278–83.
(13) Boccalon H, Marquery MC, Genestet MC, Puel P. Laser Doppler velocimetry and standardized thermal test in the exploration of Raynaud's phenomenon (abstract). *International Congress of Angiology* (Abstracts Book). 1985: 29.
(14) Kallenberg CGM, Wouda AA, Kuitert JJ, Tijssen J, Wesseling H. Nifedipine in Raynaud's phenomenon: relationship between immediate, short term and long term effects. *J Rheumatol* 1987; **14**: 284–90.
(15) Tenland T. *On laser Doppler flowmetry*. Linköpping University, 1982. Thesis.
(16) Holloway GA, Watkins DW. Laser Doppler measurement of cutaneous blood flow. *J Invest Dermatol* 1977; **69**: 306–9.
(17) Enkema L, Holloway GA, Piraino DW et al. Laser Doppler velocimetry versus heat power as indicators of skin perfusion during transcutaneous O_2 monitoring. *Clin Chem* 1981; **27**: 391–6.

(18) Johnson JM, Taylor WF, Sheperd AP, Park MK. Laser Doppler skin blood flow: comparison with plethysmography. *J Appl Physiol* 1984; **56**: 798–803.

(19) Huch R, Huch A, Lübbers DW. *Transcutaneous PO₂*. New York: Thieme Stratton, 1981.

(20) Svedman P, Holmberg J, Jacobsson S *et al*. On the relation between transcutaneous oxygen tension and skin blood flow. *Scand J Plast Reconstr Surg* 1982; **16**: 133–40.

(21) Franzeck UK, Bollinger A, Huch R, Huch A. Transcutaneous oxygen tension and capillary morphologic characteristics and density in patients with chronic venous incompetence. *Circulation* 1984; **70**: 806–11.

(22) Tonnesen KH. Transcutaneous oxygen tension in imminent foot gangrene. *Acta Anaesthesiol Scand* 1978; **68** (Suppl): 107–10.

(23) Kimbey E, Fagrell B, Björnholm M *et al*. Skin capillary abnormalities in patients with Raynaud's phenomenon. *Acta Med Scand* 1984; **215**: 127–34.

(24) Krebs R. Adverse reactions with calcium antagonists. *Hypertension* 1983; 5 (Suppl 2): 125–9.

(25) Smith CD, McKendry RJR. Controlled trial of nifedipine in the treatment of Raynaud's phenomenon. *Lancet* 1982; **ii**: 1299–301.

(26) Rodeheffer RJ, Rommer JA, Wigley F, Smith CR. Controlled double-blind trial of nifedipine in the treatment of Raynaud's phenomenon. *N Engl J Med* 1983; **308**: 880–3.

(27) Kinney EL, Nicholas GG, Gallo J, Pontoriero C, Zelis R. The treatment of severe Raynaud's phenomenon with verapamil. *J Clin Pharmacol* 1982; **22**: 74–6.

(28) White CJ, Phillips WA, Abrahams LA, Watson TD, Singleton PT. Objective benefit of nifedipine in the treatment of Raynaud's phenomenon. *Am J Med* 1986; **80**: 623–5.

(29) Nilsson H, Jonason T, Leppert J, Ringquist I. The effects of the calcium-entry blocker nifedipine on cold-induced digital vasospasm. *Acta Med Scand* 1987; **221**: 53–60.

(30) Vayssairat M, Capron L, Fiessinger JN, Mathieu J-F, Housset E. Calcium channel blockers and Raynaud's disease. *Ann Intern Med* 1981; **95**: 243.

(31) Kahan A, Amor B, Menkes C. A randomised double-blind trial of diltiazem in the treatment of Raynaud's phenomenon. *Ann Rheum Dis* 1985; **44**: 30–3.

(32) Sarkozi J, Bookman AM, Mahon W *et al*. Nifedipine in the treatment of idiopathic Raynaud's syndrome. *J Rheumatol* 1986; **13**: 331–6.

(33) Woo TY, Wong RC, Campell JP, Goldgarb MT, Voorhees JJ, Callen JP. Nifedipine in scleroderma ulcerations. *Int J Dermatol* 1984; **23**: 678–80.

(34) Finch MB, Dawson J, Johnston GD. The peripheral vascular effects of nifedipine in Raynaud's syndrome associated with scleroderma: a double blind crossover study. *Clin Rheumatol* 1986; **5**: 493–8.

(35) Wigley FM, Wise RA, Malamet R, Scott TE. Nicardipine in the treatment of Raynaud's phenomenon. *Arthritis Rheum* 1987; **30**: 281–6.

(36) Miller FW, Love LA. Prevention of predictable Raynaud's phenomenon by sublingual nifedipine. *N Engl J Med* 1987; **317**: 1476.

A role for calcium antagonists as antiatherogenic agents

D. B. Weinstein

Department of Lipid and Lipoprotein Metabolism, Sandoz Research Institute, Route 10, East Hanover, New Jersey 07936, USA

SUMMARY

A large variety of antihypertensive drugs with unique chemical structures and different mechanisms of action have been shown to reduce atherogenic lesion progression in cholesterol-fed animal models. The calcium antagonists, particularly calcium channel antagonists, have been shown to reduce the accumulation of cholesterol and arterial matrix components in cholesterol-fed rabbits. Many complex mechanisms have been proposed to account for this antiatherogenic potential of calcium antagonists. These drugs may protect against arterial cell calcium overload, as well as, by altering endothelial permeability, arterial wall relaxation factors, and smooth muscle migration and proliferation, such effects have been demonstrated in model systems. Calcium channel antagonists have been demonstrated to modify smooth muscle cell and macrophage lipoprotein uptake via receptor-mediated pathways and to alter the accumulation of cellular cholesterol esters. Recent studies have indicated that some of these effects may be relatively 'nonspecific' effects of these hydrophobic drugs which can demonstrate membrane effects which appear to be independent of their activity at voltage-operated calcium channels. The animal and cellular modelling studies have provided a database which has led to the initiation of two clinical trials which are investigating the effects of dihydropyridine calcium antagonists on coronary atherogenesis in patients with defined coronary disease. A third study (MIDAS trial) is investigating the potential changes in proliferation of early atherosclerotic lesions in carotid arteries of hypertensive patients.

INTRODUCTION

Large-scale clinical trials of antihypertensive drugs have demonstrated significant benefit of lowering blood pressure on the incidence of pressure-related complications such as congestive heart failure and stroke (1). However, it is disappointing to note that despite the strong association between hypertension and accelerated atherosclerosis, these trials have failed to show beneficial effects

The use of isradipine and other calcium antagonists in cardiovascular diseases, edited by P. A. van Zwieten, 1989; Royal Society of Medicine Services International Congress and Symposium Series No. 157, published by Royal Society of Medicine Services Limited.

MAJOR FEATURES OF ARTERIAL DISEASE:

1. Endothelial changes
2. Smooth muscle proliferation
3. Accumulation of connective tissue matrix
4. Macrophage infiltration
5. Intracellular and extracellular lipid deposition

Arterial disease scoreboard

| HYPERTENSION | 1,2,3 and 4 |
| ATHEROSCLEROSIS | 1,2,3,4 and 5 |

Figure 1. *A comparison of the major features of hypertension and atherosclerosis.*

on morbidity and mortality due to coronary artery disease and myocardial infarction (2). The relationship of hypertension to atherogenesis and associated clinical complications is poorly understood. Fig. 1 presents an outline sketch of the basic similarities of the features of both hypertensive and atherogenic disease. Despite these complex features, the ability of a number of calcium antagonists and antihypertensive agents to inhibit or delay the development of arterial lesions in experimental animal models has stimulated interest in defining whether these experimental results may have clinical applicability.

A large variety of antihypertensive drugs with unique chemical structures and different mechanisms of action have been reported to alter atherogenic lesion proliferation in the cholesterol-fed rabbit model. The mechanisms which produce these effects in experimentally-induced atherosclerosis, even in normotensive animals, are not understood. In the first study using the calcium channel antagonist nifedipine in cholesterol-fed rabbits, Henry and Bentley (3) observed only small, transient decreases in blood pressure following oral dosing and observed no significant effects on blood pressure in a second similar study performed with low doses of isradipine (4). In both studies, there was significant reduction in cholesterol accumulation in the aorta of cholesterol-fed rabbits. Blumlein *et al.* (5) and Chobanian *et al.* (6) have failed to observe reductions in atherogenesis in rabbits treated with metoprolol.

ELEMENTS LINKING HYPERTENSION AND ATHEROSCLEROSIS

Early atherosclerotic lesions are characterized by: changes in arterial endothelial permeability; migration of smooth muscle cells from the arterial media into the intima; proliferation of the 'altered' smooth muscle cells; increased synthesis by these cells of collagen, elastin and proteoglycan; and increased accumulation of lipoprotein-derived cholesterol ester by these cells, as well as by macrophages that infiltrate the arterial wall. Thus, at least three major mechanisms, *changes in endothelial permeability, smooth muscle cell proliferation and accumulation of connective tissue are major features of both hypertension and atherosclerosis.*

Campbell and Campbell (7) have described two major phenotypes of arterial smooth muscle cells. The first of these types are the normal contractile cells that make up the layers of the arterial intima. These cells are largely filled with myofilaments and are responsible for the normal contraction–relaxation cycle of the vessel wall. These cells under normal conditions maintain basal rates of connective tissue synthesis and lipoprotein uptake and metabolism. In response to continuous injury (e.g. physical perturbations, chronic hyperlipidaemia, chemical agents, immunological stress, etc.) the smooth muscle cells begin to migrate into the arterial intima where they demonstrate the second major phenotype. These altered 'synthetic' cells begin to proliferate and synthesize and secrete large amounts of extracellular matrix components. Lipoprotein uptake and intracellular lipid accumulation occurs in these altered smooth muscle cells. It is this second phenotype that is often modelled in cell culture systems using cells grown from arterial medial explants. However, it is not clear how the artificial *in vitro* environment may alter either the qualitative or quantitative responses of these differentiated cells.

The role of hyperlipidaemia in arterial disease has been examined in a variety of animal model systems. Balloon angioplasty represents one model system that rather clearly indicates the role of hyperlipidaemia in the arterial response to injury. For example, if angioplasty is performed in normolipidaemic animals, intimal smooth muscle proliferation occurs following the initial loss of endothelium and platelet deposition but intimal proliferation does not always progress (8). In animals with chronic hyperlipidaemia, angioplasty leads to extensive fibrous, cellular and lipid accumulation that most often progresses to severe narrowing of the vessel lumen and thrombosis (9). Thus, chronic hyperlipidaemia sustains or maintains the responses to injury that lead to intimal proliferation and connective tissue matrix accumulation in the diseased artery.

CALCIUM ANTAGONISTS AND ATHEROSCLEROSIS IN ANIMAL MODELS

The observation that both calcium and cholesterol accumulate in the arterial wall with increasing age has been a major factor in the design and rationale of many experiments over the last 80 years. These studies have been discussed in a number of recent reviews (10,11). Fleckenstein-Grün and co-workers (12) developed a model of vascular damage using overdoses of vitamin D_3 in rats which mimicked the histology observed in Mönckeberg's type of calcifying arteriosclerosis in humans. In these animals, the arterial medial cells are overloaded with calcium deposits and the intimal cells are heavily damaged. This model system is an excellent system for examining smooth muscle cell changes that reflect intracellular calcium overload or calcinosis. More recently, these studies have been extended to the treatment of spontaneously hypertensive rats and salt-sensitive Dahl rats with a variety of calcium channel antagonists. The calcium channel antagonists not only normalized the arterial blood pressure in these hypertensive animals but also prolonged life span and inhibited the development of arteriolar constriction and microaneurysms in the ocular fundus of aged rats (13). Normalization of blood pressure with dihydralazine (a vasodilator) under identical conditions did not have a protective effect during the 6-month study designs in hypertensive animals. However, the Fleckenstein model does not generate lipid-filled lesions within the arterial wall and therefore is not an appropriate model for evaluating the relationship between hyperlipidaemia and atherogenesis.

The cholesterol-fed rabbit model of experimental atherosclerosis has been adopted as a 'universal' test model for the determination of the antiatherogenic activity of a large variety of calcium antagonists and hypolipidaemic agents (10,11). The agents used to alter calcium accumulation or calcium metabolism included: sulphated polysaccharides including heparin (14) and chondroitin sulphate (15); EDTA (16); propranolol, reserpine and guanethidine (17); diphosphonic acid derivatives (18,19); colcemid (19); and the trivalent metal ion lanthanum (20,21). In general, all of these agents inhibited the progression of lipid-filled lesion development in the aorta of cholesterol-fed animals. The most intriguing experiments in this series were the studies performed by Kramsch and his colleagues with daily addition of 40 mg of $LaCl_3$ to the atherogenic cholesterol + fat dietary regimen (20). Lanthanum displaces calcium from membranes and is an effective antagonist for calcium binding but does not readily cross the plasma membrane. The addition of lanthanum to the atherogenic diet prevented arterial accumulation of calcium, cholesterol and matrix proteins. Unfortunately, lanthanum is not the antiatherogenic drug of choice because it is toxic in high doses. Nevertheless, it is important to attempt to understand how a simple trivalent ion that competes with calcium at membrane sites can so profoundly affect lipid accumulation and matrix protein accumulation in the artery wall.

EXPERIMENTAL STUDIES WITH CALCIUM CHANNEL ANTAGONISTS

Fleckenstein's early work on the potential protective effects of calcium channel antagonists and the increasing use of these drugs for treatment of many coronary artery disease indications, led Henry and Bentley (3) in 1981 to test whether nifedipine, a dihydropyridine calcium channel antagonist that had good hypotensive activity but no anticalcifying activity, could inhibit the progression of early arterial lesions in the cholesterol-fed rabbit model. In a placebo-controlled experiment, rabbits fed on a 2% cholesterol diet were given 20 mg of nifedipine by oral administration twice a day for 8 weeks. At the dosage used (15–17 mg/kg/day) that is approximately 20 to 40 times the normal human dose range for nifedipine, there was a 64% decrease in the aortic surface area covered by Sudan IV lipid-staining and 38% decrease in aortic cholesterol content compared with placebo-treated rabbits. After each dose of nifedipine, mean arterial pressure decreased by 10–12 mm Hg but the effects were transient and returned to baseline in less than 2 hours. This successful experiment led to a number of subsequent studies with a variety of different calcium channel antagonists. Nifedipine has been tested by several groups (22,23) at doses equal to or greater than the original dosage used by Henry and Bentley with similar qualitative results. Willis et al. (23) found that 80 mg/kg/day of nicardipine or nifedipine decreased aortic cholesterol accumulation by 69% and 75%, respectively compared with cholesterol-fed control animals. Ginsberg et al. (24) showed that large doses of diltiazem (103 mg/kg/day) reduced aortic lesion surface area by 37% but did not decrease lesion area in the intramural coronary arteries. Rouleau et al. (25) observed that 8 mg/kg/day of verapamil decreased the lipid-stained area of aortas, but the results were not statistically different from that of cholesterol-fed controls due to relatively poor bioavailability of the drug as measured by very low serum verapamil levels. In other studies with anipamil, a calcium antagonist structurally related to verapamil, Catapano et al. (26) showed that 10 mg/kg/day decreased the rabbit aortic cholesterol content by 28% compared with untreated control animals. Blumlein et al. (5) and Sievers et al. (27) extended their earlier studies with verapamil by increasing the dosage or providing the drug in the drinking water in order to overcome the bioavailability problem. In the latter study, rabbits were kept on the cholesterol dietary regimen for up to 24 weeks (27). When verapamil was maintained in the drinking water for the entire 24-week period there was 32% decrease in aortic lesion area. However, when verapamil was included for only weeks 13 to 24 of the cholesterol dietary regimen there was no significant change in aortic lesion area. The data suggests that calcium channel antagonists may be effective in early stages of atherogenic lesion formation (weeks 1 to 12 on the cholesterol dietary load) but not at later stages of lesion development in the arterial wall. This observation may also help explain the apparent paradox that nifedipine (40 mg/day) inhibits lesion progression in cholesterol-fed rabbits but does not inhibit lesion formation in the Watanabe Heritable Hyperlipidaemic (WHHL) rabbit (22). Lesion progression in the WHHL rabbit is already well-advanced before channel antagonists therapy is initiated. In the cholesterol-fed normal rabbit studies the treatment with calcium channel antagonists is started within 2 weeks of the addition of cholesterol to the diet or several weeks before visible lesions can be observed in the thoracic aorta. Blumlein et al. (5) also compared the responses of verapamil to drugs such as metoprolol (2.5 mg/kg/day) and hydralazine (2 mg/kg/day) that reduce blood pressure or cardiac contractility and found that these drugs did not reduce the extent of atherogenic lesions in the cholesterol-fed rabbit model. Thus, the negative inotropic potential of verapamil is not related to the antiatherogenic potential of this agent. They concluded that the protective effects of calcium channel antagonists were most likely due to decreases in the intracellular calcium concentrations in the arterial wall smooth muscle cells.

All of the previous studies described used large doses of channel antagonists that are between 20 to 80 times the typical human dose of these drugs. The data on bioavailability of calcium channel antagonists in rabbits suggests that plasma levels of some of the drugs within this class will be very low even at high oral dosages. Thus, many studies were designed using very high oral dosage regimens. However, if drug bioavailability were similar in rabbits and humans, it could be suggested that the very high doses of channel antagonists may have altered arterial lesion progression by pharmacological effects that are unrelated to calcium channel blockade. Therefore, Henry and colleagues repeated their original placebo-controlled experimental design with isradipine (provided by Sandoz Research Institute), a more potent dihydropyridine channel antagonist with high bioavailability that could be tested at approximately the normal human dose range. Using isradipine (PN 200–110, 0.3 mg/kg/day) at a dose only 2-fold higher than the average human dosage, they found that both cholesterol accumulation and arterial lesion surface area were reduced by 30% compared with untreated controls (4). In the same population of rabbits, isradipine was also shown to preserve the cholinergic endothelium-dependent vascular relaxation capacity of arteries that is lost upon cholesterol feeding (4). Thus, low doses of isradipine have been shown to preserve an important endothelial function and also inhibit cholesterol deposition in aortas of cholesterol-fed rabbits without altering plasma lipid levels. Table 1 presents a summary of current published experimental data showing the antiatherogenic effects of orally administered calcium channel antagonists in the cholesterol-fed rabbit model.

Effects of calcium channel antagonists on endothelial function and smooth muscle proliferation have been derived from other animal model systems. Betz (28) has developed a model for carotid artery atherogenesis that used stimulation of the arterial wall with weak electrical current to create lesions in both chow-fed or cholesterol-fed rabbits. Horse radish peroxidase injections were used to determine the permeation of macromolecules through the endothelium of anaesthetized rabbits from both the electrically stimulated and control animals. Calcium channel antagonists of several classes were shown to reduce the elevated levels of endothelial transport of the peroxidase in the electrically-stimulated

Table 1. *Published reports of the antiatherogenic effects of orally-administered calcium channel antagonists in cholesterol-fed rabbits*

Drug	Experimental dosage	Human dosage range (mg/kg/day)	Decrease in aortic: lesion area	Decrease in aortic: total cholesterol
Nifedipine	15–17 mg/kg/day	0.43–0.86	64%	38%
	40 mg/day		53%	
	20 mg/day		58%	69%
Nicardipine	80 mg/kg/day	1–2	49%	75%
Diltiazem	103 mg/kg/day	2.6–5.2	37%	—
Verapamil	50–200 mg/kg/day	3.7–7.4	47%	—
	2 grams/l in drinking water		32%	—
Anipamil	10 mg/kg/day	—	25%	28%
Israidipine	*0.3 mg/kg/day*	*0.07–0.14*	31%	29%

arteries. The data suggested that both vesicular transport through cells and transport through intercellular clefts were reduced. In this same model, smooth muscle cell proliferation was inhibited in a dose-dependent manner by 1–30 mg/kg of flunarizine but oral administration of verapamil and nimodipine were ineffective due to low bioavailability in these studies (28).

Another experimental model of interest in rats and rabbits utilizes balloon catheter removal of the entire endothelial cell barrier in the carotid artery. This procedure causes a very rapid proliferative response resulting largely from platelet-derived growth factors released after platelet adhesion to the subendothelial matrix of the damaged vessels. Although endothelial desquamation and platelet deposition are not believed to be major initiating mechanisms in the typical human arterial lesion in major arteries, these models have provided interesting data. Handley *et al.* (29) showed that within 2 weeks of arterial damage by balloon catherization (in the absence of hyperlipidaemia) in rats there was a progressive migration and cell proliferative response that formed extensive, reproducible fibromuscular lesions in the carotid arteries. When two Sandoz calcium channel antagonists (PY 108–068 and isradipine) were given subcutaneously at 1 mg/kg/day during the 2 weeks after endothelial denudation, there was a reduction of 25% and 44%, respectively, in the intimal lesion cross-sectional area, and significant reductions in matrix accumulation were noted. In similarly designed studies, Jackson *et al.* (30) measured ^{3}H-thymidine incorporation into rat aortic DNA 48 hours after balloon injury. Oral administration of nifedipine (2 mg/kg), verapamil (100 mg/kg), and diltiazem (100 mg/kg) significantly reduced arterial DNA synthesis but had no effect on DNA synthesis in proliferating tissues such as bone marrow, testes, and duodenum. In balloon catheterized rabbits, 20 mg/kg/day of nifedipine reduced neointimal cross-sectional area by 39% at 14 days after injury. The data suggest that in these models of rapid smooth muscle cell proliferation, large doses of channel antagonists are associated with antiproliferative activities.

POTENTIAL ANTIATHEROGENIC MECHANISMS

Calcium channel antagonists do not significantly alter cholesterol or lipoprotein levels in plasma and therefore have limited effects on experimentally-induced hyperlipidaemia in either animal models of atherogenesis or in human studies (31–33). In order to explore other potential mechanisms by which calcium antagonists may alter the progression of atherogenesis, one is forced to leave the complex, multifactorial disease state itself and seek answers in simpler systems such as cell culture systems using smooth muscle cells and macrophage-like cells as models of arterial wall metabolism. Many calcium-regulated processes that occur at the arterial wall have been proposed as potential sites for interaction of calcium antagonists (15,34). If one considers early events in lesion formation as potential targets of calcium channel antagonists the five events that may be most important for controlling lesion proliferation would be: 1) prevent metabolic changes leading to disruption of the endothelial barrier; 2) inhibit migration of smooth muscle cells into the arterial intima and inhibition of the migration of neutrophils and macrophages into the arterial wall; 3) inhibit the proliferation of smooth muscle cells in the arterial intima; 4) inhibit the increased accumulation of extracellular matrix; and 5) inhibit lipid accumulation by smooth muscle or macrophage-derived 'foam cells' in the artery wall.

In cell culture systems, calcium channel antagonists have been shown to inhibit the migration response of cells to chemotactic agents such as formyl-methionyl-leucyl-phenylalanine (35). Calcium channel antagonists have been shown to inhibit the proliferation of smooth muscle cells isolated from atherosclerotic segments of human aorta (36) although these effects required verapamil concentrations $>10^{-5}$M. However, Stein et al. (37) have shown that 50 μM verapamil significantly inhibited the accelerated DNA synthesis and cell proliferation that occurred when cells were grown for long periods in the presence of cholesterol ester-rich lipoprotein fractions from the serum of hypercholesterolaemic rabbits.

Abnormal accumulation of extracellular connective tissue in blood vessels is a hallmark of the pathology associated with both hypertension and atherogenesis. It is currently believed that the extracellular matrix is a major regulator of cell repair following injury and re-endothelialization of vascular injury sites is, in part, dependent upon both the composition and organizational state of the sub-endothelial extracellular matrix (38). It has recently been demonstrated that the synthesis rate of collagen, as determined in vivo following an intravenous bolus of ^{3}H-proline, is increased ten-fold in the intima plus inner media of atherosclerotic thoracic arteries from rats fed atherogenic diets compared with arteries from animals fed on normal chow diets (39). Similarly, it has been shown that elastin synthesis is two- to four-fold elevated in pulmonary artery tissue and medial smooth muscle cells from animals with pulmonary hypertension compared with controls (40). Ingber and Folkman (41) have demonstrated that heparin and angiogenic steroids induce basement membrane breakdown which leads to inhibition of proliferation of capillary endothelial cells. These authors have also demonstrated that inhibitors of collagen synthesis and crosslinking also block capillary endothelial proliferation. The interactions between matrix proteins or glycoproteins and migration and proliferation of vascular smooth muscle cells in normal repair mechanisms and in abnormal pathological states are obviously complex. A second major factor to be considered must be the well-characterized ability of the arterial extravascular matrix (elastin and proteoglycans) to bind and retain significant quantities of low density lipoprotein (42). Thus, any safe therapeutic agent which might reduce the synthesis of matrix components in the vessel wall of hypertensives or at the sites of atherogenic lesions might have profound effects on vascular cell migration and proliferation. Our laboratory has previously demonstrated that isradipine and some, but not all, other calcium channel antagonists have the potential to inhibit the incorporation of both

[14]C-proline into hydroxyproline-containing matrix proteins and ^{35}S-sulphate into sulphated proteoglycans by vascular smooth muscle cell cultures at concentrations in the range of 1–50 μM (43). Fig. 2 shows a typical experiment which demonstrates the inhibition of matrix protein production in cultures of monkey arterial smooth muscle cells which were exposed to 10–50 μM concentration of either verapamil or isradipine (PN200–110) for 24 hours. The data show that isradipine, but not verapamil, significantly decreases matrix protein production under the experimental conditions. Thus, if the concentration of some channel antagonists *in vivo* can approach the micromolar level within vascular smooth muscle cells, it is possible that alteration of matrix synthesis could represent a component of the antiatherogenic activities which can be demonstrated in experimental models.

Although clinical studies show minimal effects of calcium channel antagonists on plasma lipoprotein concentrations in hypertensive humans, the interactions of calcium channel antagonists with lipoprotein metabolic pathways should not be ignored as a potential mechanism for altering cholesterol accumulation within arterial cells in the atherogenic state. Stein *et al.* (44) have reported that verapamil (10–50 μM) increased receptor-mediated clearance of LDL from cell surfaces but also caused a delay in lysosomal degradation of LDL. Similar concentrations of verapamil and diltiazem have been shown to increase the synthesis of LDL receptors in human skin fibroblasts as measured by the incorporation of labelled amino acids into immunoprecipitable LDL receptor protein (45). Although these results are of significant interest it is difficult to relate these events in skin fibroblasts to metabolic activities in cells that exist in the atherogenic arterial wall. Etingin and Hajjar (46) used smooth muscle cells from lesion and non-lesion areas of rabbit aorta that were maintained in culture for 7 days with good retention

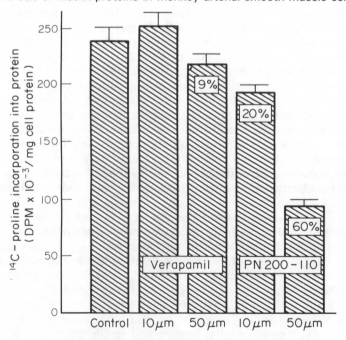

Figure 2. *The effect of calcium channel antagonists on* [14]*C-proline incorporation into matrix proteins synthesized by cultures of monkey arterial smooth muscle cells.*

of the differences in lipid content of the cells from the two different sources. Nifedipine (0.3 μM) doubled the activity of the lysosomal cholesterol ester hydrolase in the cells from the lesion areas and decreased cell cholesterol and cholesterol ester content but did not alter the enzyme activity of cells or cholesterol content from non-lesion areas. The metabolic effect of nifedipine treatment may be due to increased levels of cAMP because adenylate cyclase inhibitors blocked the induction of cholesterol ester hydrolase activity. Thus, the metabolic state of the cell in the artery wall may play a determining factor in the effects observed in the presence of calcium antagonists.

The macrophage represents a cell type that can significantly contribute to the cholesterol ester accumulation in the artery wall. Daugherty et al. (47) used cholesterol ester-rich, beta-migrating very low density lipoproteins (beta-VLDL) from cholesterol-fed rabbits to stimulate cholesterol ester deposition in macrophages isolated from normolipidaemic rabbits. Nifedipine (a Class I calcium channel antagonist) at 10^{-7}M concentration significantly inhibited cholesterol ester accumulation. However, Bay K 8644, a calcium channel agonist (related in structure to nifedipine) that increases cytoplasmic calcium concentrations, showed the same response of inhibition of cholesterol ester accumulation at equivalent drug concentrations. This experimental data is of significant interest since it is relevant to the mechanism(s) by which drugs of these types may work. Since both antagonist (nifedipine) and agonist (Bay K 8644) have the same effect on lipid accumulation in the macrophage, it may be concluded that the effects of these two dihydropyridines must be due to mechanisms other than calcium transport across voltage-operated calcium channels in this cell type. In the same experiments, verapamil (a Class II calcium channel antagonist) also reduced cholesterol ester accumulation, whereas diltiazem had no effect at any concentration tested. Recent studies by Stein et al. (48) using established macrophage-like cell lines have indicated that 24-hour treatment of cells with verapamil results in a reduction of cholesterol esterification mediated by inhibition of the enzyme acyl CoA:cholesterol acyltransferase (ACAT) and not by alterations in intracellular cholesterol transport. Fig. 3 presents a summary of three of the major sites in the cellular metabolic pathways of cholesterol accumulation and utilization which may be altered by calcium antagonists.

The observations that calcium channel antagonists may alter metabolic activities of smooth muscle cells and macrophages in a cell culture environment have been derived from experimental designs that use 10 to 500 times greater concentrations than the upper limit of these drugs in plasma. This has been of great concern to experimentalists in this field. However, it is necessary to consider whether total plasma drug concentration or the tissue bioavailability of drug is the relevant factor when correlating efficacy with drug treatment. Pang and Sperelakis (49) have

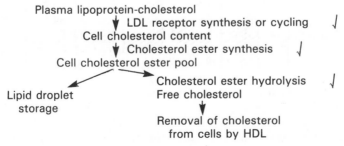

Figure 3. *Potential sites of action of calcium antagonists on cell cholesterol content.*

demonstrated that the rate of uptake of a variety of calcium channel antagonists into skeletal muscle cells is well-correlated with the organic/aqueous partition coefficients or hydrophobicity of each compound. Calcium channel antagonists are, in general, hydrophobic drugs that are carried in plasma on lipoprotein particles (50) as well as other plasma proteins. Because calcium channel antagonists are carried by lipoprotein particles in plasma, it is possible that these drugs may be taken up by arterial wall cells via both receptor-mediated and nonreceptor-mediated lipoprotein pathways or via direct uptake of the lipid core of lipoprotein particles, in addition to direct interactions with binding sites at or within voltage-operated calcium channels.

It is not possible to determine the true membrane concentration of hydrophobic drugs that may intercalate into the lipid bilayer. However, it can be speculated that local membrane concentrations of hydrophobic calcium antagonists could greatly exceed the concentration of these agents in the plasma and extracellular space. Chester et al. (51) have characterized the partitioning of dihydropyridine calcium channel antagonists into membranes. Their data suggests that these chemical agents reside within the lipid bilayer and distribute within the membrane by a process of lateral diffusion. Shi and Tien (52) have reported an interesting observation with reference to calcium channel antagonists and membrane structure. Using electron-spin resonance probes to examine the structural features and fluidity of model phospholipid membranes, they found that both verapamil and diltiazem caused large perturbations in membrane bilayer structure while a beta-blocker (sotalol) caused no changes. Thus, some of the metabolic properties of calcium channel antagonists may be due to the chemical properties of these agents that alter membrane fluidity or conformation at specific subcellular membrane sites.

CONCLUSION

The rate-limiting steps in arterial lesion progression are not well understood at this time. Animal model studies have focused on events that may be responsible for the *early* events of proliferative lesion formation. Whereas the role of hyperlipidaemia and lipid accumulation have been extensively implicated in atherogenesis it is clear that arterial Ca^{++} overload is also a major factor (12,13). Strickberger et al. (53) have used new techniques based upon kinetic modelling of ^{45}Ca efflux experiments in arterial segments and have determined that cholesterol feeding does not significantly alter the extracellular calcium content of rabbit aortas whereas intracellular aortic calcium was increased approximately five-fold compared with chow-fed controls. These experimental data support the hypothesis that increased levels of intracellular calcium may be involved in the metabolic alterations that occur within smooth muscle cells in the developing lesion.

The potential regulation of multiple pathways or mechanisms of cellular function by calcium channel antagonists raises the issue of 'specificity of action' of these drugs in these model systems. In the past several years there have been many reports which support the idea that calcium channel antagonists are 'promiscuous' receptor antagonists (54) and are capable of interacting with benzodiazepine binding sites (55), muscarinic and alpha- and beta-adrenergic receptors (56), serotonin receptors (57), and calmodulin (58). Are all these effects related to control of intracellular Ca^{++} concentration or to perturbations of membrane structure due to hydrophobic drug transport of accumulation in membranes? Many reports in the literature suggest that calcium channel antagonists may function *independently* of their effects on calcium flux. In addition to the effects described above on cholesterol esterification (47), the effect of calcium channel antagonists on matrix production also appears to be independent of calcium flux changes across voltage-operated calcium channels since this inhibitory activity has been demonstrated in cells which do not have voltage-regulated calcium channels (43). It has been reported that a variety of calcium channel antagonists can inhibit muscle contraction by mechanisms independent of their channel gating properties (59,60). Pennington *et al.* (61) reported similar findings in a neutrophil chemotaxis assay.

On the basis of the data from experimental models of atherosclerosis and from the studies derived from *ex vivo* arterial tissue studies and *in vitro* cell culture models, it is clear that calcium channel antagonists have profound effects on the maintenance of normal arterial wall function and on the progression of atherogenic lesions in animal models. Calcium channel antagonists may effectively protect the arterial wall via two independent sets of mechanisms. First, by regulating the intracellular Ca^{++} flux across voltage-operated channels they may protect the arterial wall cells against calcium overload. Secondly, calcium channel antagonists may alter a variety of metabolic activities of endothelial and smooth muscle cells that determine the rate of accumulation of components of the atherogenic lesion. It is not at all certain that atherogenic lesion formation in high-risk patients with established cardiovascular disease will be responsive to calcium antagonist therapy. Nevertheless, the *in vitro* and *in vivo* model systems have directed our attention to the possible utility of safe and effective calcium antagonists as therapeutic agents in atherosclerotic disease.

The rationale provided by these studies needs to be extended to the study of lesion proliferation in human populations at high risk for cardiovascular disease. Recently, two clinical studies have been initiated to establish whether dihydropyridine calcium channel antagonists are antiatherogenic in human subjects with established coronary artery disease. The International Nifedipine Trial on Atherosclerotic Coronary Therapy (INTACT) is the first double-blind, randomized prospective study of this type and will determine the effects of 80 mg/day of nifedipine *vs* placebo as determined by coronary angiography over a 3-year period in 426 patients with mild to moderate coronary disease (62). The second study which is underway at the Montreal Heart Institute is a relatively similar study which will utilize angiographic endpoints in 383 patients randomized to 20–30 mg/day of nicardipine or placebo (63).

The rationale for expectations of significant effects of calcium channel antagonists on pre-existing human coronary atherogenic disease in the two trials described above is not clear. In 2 or 3 year study designs, calcium channel antagonists are not likely to significantly influence regression of lipid or fibrous lesions, if indeed these agents can influence regression at all. A very different approach has been initiated by the Sandoz Research Institute which in 1988 has initiated a study designed to quantify the progression of *early lesions* in the carotid arteries of hypertensive subjects at risk for development of atherogenic disease. The Multicenter Isradipine/Diuretic Atherosclerosis Study (MIDAS) is the first double-blind, randomized trial which will evaluate early lesion progression in carotid arteries by non-invasive, B-mode ultrasound analysis. A number of recent studies have shown the utility of real-time (B-mode) ultrasound analysis to measure changes in carotid arterial wall thickness (64,65). The MIDAS study will include a minimum of 600 hypertensive patients who will be titrated to equivalent blood pressure reduction with isradipine or conventional diuretic therapy. Based upon the known rate of carotid lesion proliferation it will be most challenging to determine whether a calcium antagonist can alter the rate of atherogenesis in the carotid artery. Such studies in clinical trials can be expected to have a great impact on future therapy.

REFERENCES

(1) Julius S. Managing the cardiovascular risks in hypertension. *Am Heart J* 1988; **116** 265–6.
(2) Chobanian A. Atherosclerosis: the hypertension connection? *Am Heart J* 1988; **116**: 319–22.
(3) Henry PD, Bentley KI. Suppression of atherogenesis in cholesterol-fed rabbits treated with nifedipine. *J Clin Invest* 1981; **68**: 1366–9.
(4) Habib JB, Bosaller C, Wells S, Williams C, Morrissett JD, Henry PD. Preservation of endothelium-dependent vascular relaxation in cholesterol-fed rabbits by treatment with the calcium channel blocker PN 200–110. *Circ Res* 1986; **58**: 305–9.
(5) Blumlein SL, Sievers R, Kidd P, Parmley WW. Mechanisms of protection from atherosclerosis by verapamil in the cholesterol-fed rabbit. *Am J Cardiol* 1984; **54**: 884–9.
(6) Chobanian AV. Effects of calcium channel antagonists and other antihypertensive drugs on atherogenesis. *J Hypertension* 1987; 5(Suppl 4): S43–8.
(7) Campbell GR, Campbell JH. Recent advances in molecular pathology: smooth muscle phenotypic changes in arterial wall homeostasis. Implications for the pathogenesis of atherosclerosis. *Exp Mol Pathol* 1985; **42**: 139–62.
(8) Wolinsky H. Insights into coronary angioplasty-induced restenosis from examination of atherogenesis. *Am J Cardiol* 1987; **60**: 65B–7B.
(9) Faxon DP, Sanborn TA, Weber VJ, Haudenschild C, Gottsman SB, McGovern WA, Ryan TJ. Restenosis following transluminal angioplasty in experimental atherosclerosis. *Arteriosclerosis* 1984; **4**: 189–95.
(10) Parmley WW. Calcium channel blockers and atherogenesis. *Am J Med* 1987; **82**: 3–8.
(11) Weinstein DB, Heider JG. Antiatherogenic properties of calcium channel blockers. *Am J Med* 1988; **84**: 102–8.
(12) Fleckenstein-Grün G, Frey M, Fleckenstein A. Calcium antagonists: mechanisms and therapeutic uses. *Trends Pharm Sci* 1984; **5**: 283–6.
(13) Fleckenstein A, Fleckenstein-Grün G, Frey M, Zorn J. Future directions in the use of calcium antagonists. *Am J Cardiol* 1987; **59**: 177B–87B.
(14) Gutmann N, Constantinides P. Inhibition of experimental atherosclerosis by sulfated alginic acid. *Arch Pathol* 1955; **59**: 717–22.
(15) Morrison LM, Bajwa GS, Alfin-Slater RB, Ershoff BH. Prevention of vascular lesions by chondriotin sulfate A in the coronary artery and aorta of rats induced by a hypervitaminosis D, cholesterol-containing diet. *Atherosclerosis* 1972; **16**: 105–18.
(16) Wartman A, Lampe TL, McDann DS, Boyle AJ. Plaque reversal with Mg EDTA in experimental atherosclerosis: elastin and collagen metabolism. *J Atheroscler Res* 1967; **7**: 331–41.
(17) Whittington-Coleman PJ, Carrier O Jr, Douglas BH. The effects of propranolol on cholesterol-induced atheromatous lesions. *Atherosclerosis* 1973; **18**: 337–45.
(18) Potokar M, Schmidt-Dunker M. The inhibitory effect of new diphosphonic acids on aortic and kidney calcification in vivo. *Atherosclerosis* 1978; **30**: 313–20.
(19) Kramsch DM, Chan CT. The effects of agents interfering with soft tissue calcification and cell proliferation on calcific fibrous-fatty plaques in rabbits. *Circ Res* 1978; **42**: 562–72.
(20) Kramsch DM, Aspen AJ, Apstein CS. Suppression of experimental atherosclerosis by the Ca^{++}-antagonist lanthanum. *J Clin Invest* 1980; **65**: 967–81.
(21) Kramsch DM, Aspen AJ, Rozier LJ. Atherosclerosis: prevention by agents not affecting abnormal levels of blood lipids. *Science* 1981; **213**: 1511–2.
(22) Watanabe N, Ishikawa Y, Okamoto R, Watanabe Y, Fukuzaki H. Nifedipine suppressed atherosclerosis in cholesterol-fed rabbits but not in Watanabe Heritable Hyperlipidemic rabbits. *Artery* 1987; **14**: 283–94.
(23) Willis A, Nagel B, Churchill V, Whyte MA, Smith DL, Mahmud L, Puppione DL.

Antiatherosclerotic effects of nicardipine and nifedipine in cholesterol-fed rabbits. *Arteriosclerosis* 1985; **5**: 250–5.

(24) Ginsberg R, David K, Bristow MR, McKennel K, Billingham ME, Schroeder JS. Calcium antagonists suppress atherogenesis in aorta but not in intramural coronary arteries of cholesterol-fed rabbits. *Lab Invest* 1983; **49**: 154–8.

(25) Rouleau J-L, Parmley WW, Stevens J, Wilkman-Coffelt J, Sievers R, Mahley RM, Havel RJ. Verapamil suppresses atherosclerosis in cholesterol-fed rabbits. *J Am Clin Chem* 1983; **1**: 1453–60.

(26) Catapano AL, Maggi FM, Cicerano U. The antiatherosclerotic effect of anipamil in cholesterol-fed rabbits. *Ann NY Acad Sci* 1988; **522** 519–22.

(27) Sievers R, Rashid T, Garrett J, Blumlein S, Parmley WW. Verapamil and diet halt progression of atherosclerosis in cholesterol-fed rabbits. *Cardiovasc Drugs Ther* 1987; **1**: 65–9.

(28) Betz E. The effect of calcium antagonists on intimal cell proliferation in atherogenesis. *Ann NY Acad Sci* 1988; **522**: 399–410.

(29) Handley DA, Van Valen RG, Melden MK, Saunders RN. Suppression of rat carotid lesion development by the calcium channel blocker PN 200–110. *Am J Pathol* 1986; **124**: 88–93.

(30) Jackson CL, Bush RC, Bowyer DE. Inhibitory effect of calcium channel antagonists on balloon-induced arterial smooth muscle cell proliferation and lesion size. *Atherosclerosis* 1988; **69**: 115–22.

(31) Pool PE, Seagren SC, Salel AF, Skalland ML. Effects of diltiazem on serum lipids, exercise performance and blood pressure: randomized, double-blind, placebo-controlled evaluation for systemic hypertension. *Am J Cardiol* 1985; **56**: 86H–91H.

(32) Sasaki N, Matsuoka N, Saito Y, Yoshida S. The effect of nifedipine on serum lipoproteins in hyperlipoproteinemic subjects. *Curr Ther Res* 1988; **43**: 317–26.

(33) Rauraama R, Taskinen E, Seppänen K, Rissanen V, Salonen R, Venäläinen JM, Salonen JT. Effects of calcium antagonist treatment on blood pressure, lipoproteins, and prostaglandins. *Am J Med* 1988; **84**: 93–6.

(34) Henry PD. Calcium antagonists as antiatherogenic agents. *Ann NY Acad Sci* 1988; **522**: 411–9.

(35) Williamson KC, Tauber AI, Navarro J. Nisoldipine inhibits formyl-methionyl-leucyl-phenylalanine receptor-coupled calcium transport in human neutrophils. *J Leukocyte Biol* 1987; **42**: 239–44.

(36) Orekhov AN, Tertov VV, Khashimov KA, Kudryashov SS, Smirnov VN. Evidence of antiatherosclerotic action of verapamil from direct effects on arterial cells. *Am J Cardiol* 1987; **59**: 495–6.

(37) Stein O, Halperin G, Stein Y. Long-term effects of verapamil on aortic smooth muscle cells cultured in the presence of hypercholesterolemic serum. *Arteriosclerosis* 1987; **7**: 585–92.

(38) Madri JA, Pratt BM, Yannariello-Brown J. Matrix-driven cell size change modulates aortic endothelial cell proliferation and sheet migration. *Am J Pathol* 1988; **132**: 18–27.

(39) Opsahl WP, DeLuca DJ, Erhart LA. Accelerated rates of collagen synthesis in atherosclerotic arteries quantified in vivo. *Arteriosclerosis* 1987; **7**: 470–6.

(40) Mecham RP, Whitehouse LA, Wrenn DS, *et al.* Smooth muscle-mediated connective tissue remodelling in pulmonary hypertension. *Science* 1987; **237**: 423–6.

(41) Ingber D, Folkman J. Inhibition of angiogenesis through modulation of collagen metabolism. *Lab Invest* 1988; **59**: 44–51.

(42) Srinivasan SR, Vijayagopal P, Dalferes ER, Abbate B, Radhakrishnamurthy B, Berenson GS. Low density lipoprotein retention by aortic tissue. Contribution of extracellular matrix. *Atherosclerosis* 1986; **62**: 201–8.

(43) Heider JG, Weinstein DB, Pickens CE, Lan S, Su C-M. Anti-atherogenic activity of the calcium channel blocker isradipine (PN 200–110): a novel effect on matrix synthesis independent of calcium channel blockade. *Transpl Proc* 1987; **29**(Suppl 5): 96–101.

(44) Stein O, Leitersdorf E, Stein Y. Verapamil enhances receptor-mediated endocytosis of low density lipoproteins by aortic cells in culture. *Arteriosclerosis* 1985; **5**: 35–44.

(45) Filipovic I, Buddecke E. Calcium channel blockers stimulate LDL receptor synthesis in human skin fibroblasts. *Biochem Biophys Res Commun* 1986; **136**: 845–50.

(46) Etingin OR, Hajjar DJ. Nifedipine increases cholesterol ester hydrolytic activity in lipid-laden arterial smooth muscle cells. *J Clin Invest* 1985; **75**: 1554–8.

(47) Daugherty A, Rateri DL, Schonfeld G, Sobel BE. Inhibition of cholesterol ester deposition in macrophages by calcium entry blockers: an effect dissociable from calcium entry blockade. *Br J Pharmacol* 1987; **91**: 113–8.

(48) Stein O, Stein Y. Effect of verapamil on cholesterol ester hydrolysis and reesterification in macrophages. *Arteriosclerosis* 1987; **7**: 578–84.

(49) Pang DC, Sperelakis N. Uptake of calcium antagonistic drugs into muscle as related to their lipid solubilities. *Biochem Pharmacol* 1984; **33**: 821–6.

(50) Kwong TC, Sparks JD, Sparks CE. Lipoprotein and protein binding of the channel blocker diltiazem. *Proc Soc Exp Biol Med* 1985; **178**: 313–6.

(51) Chester DW, Herbette LG, Mason RP, Joslyn AF, Triggle DJ, Koppel DE. Diffusion of dihydropyridine calcium channel antagonists in cardiac sarcolemmal lipid multibilayers. *Biophys J* **1987**; **52**: 1021–30.

(52) Shi B, Tien HT. Action of calcium channel and beta-adrenergic blocking agents in bilayer lipid membranes. *Biochem Biophys Acta* 1986; **859**: 123–34.

(53) Strickberger SA, Russek LN, Phair RD. Evidence for increased aortic plasma membrane calcium transport caused by experimental atherosclerosis in rabbits. *Circ Res* 1988; **62**: 75–80.

(54) Insel PA. Phenylalkylamines are promiscuous receptor blockers. *Trends Pharmacol Res* 1988; **9**: 34.

(55) Cantor EH, Kennessey A, Semenuk G, Spector S. Interaction of calcium channel blockers with non-neural benzodiazepine binding sites. *Proc Natl Acad Sci USA* 1984; **81**: 1549–52.

(56) Thayer SA, Welcome M, Chabra A, Fairhurst AS. Effects of dihydropyridine calcium channel blocking drugs on rat brain muscarinic and α-adrenergic receptors. *Biochem Pharmacol* 1985; **34**: 175–80.

(57) Ohashi M, Kanai R, Takayangi I. Do D-600 and diltiazem interact with serotonin receptors in rabbit vascular tissues. *J Pharmacol Exp Ther* 1985; **233**: 830–5.

(58) Walsh MP, Sutherland C, Scott-Woo GC. Effects of felodipine (a dihydropyridine calcium channel blocker) and analogues on calmodulin-dependent enzymes. *Biochem Pharmacol* 1988; **37**: 1569–80.

(59) Raddino R, Poli E, Ferrari R, Visioli O. Effects of calcium entry blockers not connected with calcium channel inhibition. *Gen Pharmacol* 1987; **18**: 431–6.

(60) Jacquemond V, Rougier O. Nifedipine and Bay K inhibit contraction independently from their action of calcium channels. *Biochem Biophys Res Commun* 1988; **152**: 1002–7.

(61) Pennington JE, Kemmerich B, Kazanjian PH, Marsh JD, Boerth LW. Verapamil impairs human neutrophil chemotaxis by a non-calcium-mediated mechanism. *J Lab Clin Med* 1986; **108**: 44–52.

(62) Lichtlen PR, Nellesen U, Rafflenbeul W, Jost S, Hecker H. International trial on antiatherosclerotic therapy (INTACT). *Cardiovasc Drugs Therapy* 1987; **1**: 71–9.

(63) Waters D, Freedman D, Lesperance J *et al*. Design features of a controlled clinical trial to assess the effect of a calcium entry blocker upon the progression of coronary artery disease. *Contr Clin Trials* 1987; **8**: 216–42.

(64) Poli P, Tremoli E, Colombo A, Sirtori M, Pignoli P, Paoletti R. Ultrasonographic measurement of the common carotid artery wall thickness on hypercholesterolemic patients. *Atherosclerosis* 1988; **70**: 253–61.

(65) Ricotta JJ, Bryan FA, Bond MG *et al*. Multicenter validation study of real-time (B-mode) ultrasound, ateriography and pathologic examination. *J Vasc Surg* 1987; **6**: 512–20.

Conclusions

P. A. van Zwieten

Departments of Pharmacotherapy and Cardiology, Academic Medical Centre,
University of Amsterdam, The Netherlands

1. In the course of the past decade considerable clinical experience has been obtained with the dihydropyridines and other calcium antagonists in the treatment of hypertension and various forms of angina pectoris, whereas limited data have become available concerning the treatment of other cardiovascular disorders, like for instance hypertrophic cardiomyopathy. Simultaneously, a deepened insight into the mode of action of calcium antagonists has been obtained, thus improving as well our fundamental knowledge of calcium metabolism in vascular smooth muscle, cardiac and other tissues. It was the aim of the round table conference held in Amsterdam on January 20, 1989 to critically discuss the newer aspects of calcium antagonists from both fundamental and clinical viewpoints, and to position the new compound *isradipine* in comparison with the so far developed dihydropyridines.

Isradipine (PN 200-110, Lomir®) is a calcium antagonist which chemically belongs to the major subgroup of the dihydropyridines of which nifedipine and nitrendipine are the prototypes. Its pharmacological and haemodynamic profiles indeed largely correspond to that of the dihydropyridines. However, in animal models isradipine has been claimed to show a stronger inhibitory activity on the development of atherosclerotic plaques than that observed for other calcium antagonists. The relevance of this finding in humans remains to be demonstrated.

With respect to the newer developments in the field of calcium antagonists in a more general perspective, particular attention was paid to the influence of these drugs on vascular smooth muscle, to their obvious and well-known tissue specificity (which is rather different for the various compounds developed) as well as to their ancillary properties, like potential anti-atherogenic and cytoprotective potencies.

The experiments described by Struyker Boudier and co-workers have provided new and unexpected information concerning the influence of calcium antagonists on the microcirculation, thus showing that *regional myogenic tone* is a most important basic parameter in determining the vasodilator response to calcium antagonists. Since myogenic tone largely differs in various types of blood vessels and also in different species, these findings may satisfactorily explain the obvious tissue specificity of the calcium antagonists belonging to the well-known subgroups.

The use of isradipine and other calcium antagonists in cardiovascular diseases, edited by P. A. van Zwieten, 1989; Royal Society of Medicine Services International Congress and Symposium Series No. 157, published by Royal Society of Medicine Services Limited.

The vasodilator potency of the calcium antagonists, especially that of the dihydropyridines at the level of the microcirculation also offers a plausible explanation of the oedema in the legs, a well-known adverse reaction, which does *not* reflect fluid retention as a result of general systemic vasodilatation.

Finally, the influence of calcium antagonists on resistance vessels involves different aspects in acute and long-term treatment with these drugs: in the *acute* situation both myogenic tone and nervous control of vascular activity are influenced, whereas during *prolonged* treatment not only vascular tone, but also other factors are modulated, like for instance the severity of structural vascular changes associated with long existing hypertension and also the density of blood vessels in the microcirculation.

2. Apart from the general profile of isradipine as mentioned above recent animal experiments presented by Hof and co-workers may reveal a few potentially different properties of this compound in comparison with the existing dihydropyridines.

Accordingly, isradipine offers functional and biochemical protection against cerebral ischaemia, both as pre- and post-event treatment in an animal model of acute stroke. The dose-dependent effect could be characterized by a bell-shaped dose-response curve, suggesting a rather narrow therapeutic range. Few data for other calcium antagonists are available, but nimodipine proved less effective than isradipine.

A further interesting observation was the much stronger anti-vasoconstrictor activity of isradipine in atherosclerotic rabbit vessels in comparison to that in preparations from normal animals, although this effect is probably not specific for isradipine.

Finally, in an open chest rabbit preparation isradipine displays a significantly weaker negative inotropic effect than nifedipine, thus confirming the well-known vascular specificity of isradipine.

3. The clinical application of calcium antagonists in *hypertension*, as reviewed by Man in 't Veld may be resumed as follows: The haemodynamic profile of the calcium antagonists, characterized by arterial vasodilatation and a reduction of total peripheral resistance readily explains the antihypertensive potency of these drugs. The transient rise in heart rate, characteristic for the dihydropyridines is caused by reflex activation of the sympathetic system. The depressant influence of verapamil and diltiazem on sinus node activity and A-V conduction prevents a rise in heart rate during treatment with these and related compounds. A further interesting property of the dihydropyridines is their modest natriuretic potency, which prevents fluid retention in spite of potent vasodilatation. This natriuretic potency cannot be explained by renal vasodilatation alone. A direct tubular effect has been proposed, as well as an enhanced release of the atrial natriuretic factor (ANF) during treatment with dihydropyridines. These drugs also enhance plasma renin activity (PRA), but this rise in PRA is for so far unknown reasons not followed by a concomitant elevation of plasma aldosterone levels.

Calcium antagonists can be applied in all types of hypertension and in various subgroups of patients, including the elderly, black and low-renin hypertensives. Although suggested by some authors the preferential use of calcium antagonists in elderly patients is not substantiated by the available clinical data. Most of the side-effects of calcium antagonists reflect vasodilatation (headache, flush) or intestinal smooth muscle relaxation (obstipation). As discussed previously the oedema in the extremities is now satisfactorily explained at the level of the microcirculation.

Calcium antagonists can be combined with various other antihypertensives, like diuretics, beta-blockers (in particular dihydropyridines, to be combined with beta-blockers without ISA), and ACE-inhibitors. A combination of drugs is even recommended in many cases and appears to deserve further attention in clinical and basic research.

Calcium antagonists in the management of hypertension may offer advantages like the reduction of left ventricular and vascular hypertrophy and prevention or even regression of atherosclerotic and other structural changes in blood vessels, but such phenomena remain to be demonstrated under clinical conditions and on a larger scale. If indeed effective in this sense, the calcium antagonists might be anticipated to be more active in reducing the incidence and severity of coronary heart disease associated with hypertension than so far demonstrated for beta-blockers and diuretics.

As demonstrated by De Bruijn, isradipine in doses of 5 or 10 mg b.i.d., administered as such or as a slow release preparation proved an effective antihypertensive, associated with very little or no rise in heart rate (slow release preparation). Adequate blood pressure control is obtained with a satisfactory responder rate and without disturbing the circadian rhythm of blood pressure. The side-effects were those to be expected for a dihydropyridine. The natriuretic effect was also demonstrated for isradipine and appeared to be maintained throughout a period of two years of treatment. The clearance of uric acid proved somewhat enhanced by isradipine, although the clinical relevance of this finding remains unclear so far.

4. In *cardiology*, as discussed by Lie, the dihydropyridines are mainly useful in the management of stable and Prinzmetal's angina, and possibly in unstable angina although less convincingly so. Their usefulness in heart failure is subject to debate.

The therapeutic efficacy of dihydropyridine-calcium antagonists in stable angina is largely determined by their haemodynamic activities, involving coronary dilatation, the relief of coronary spasm (if present), and a reduction in cardiac afterload as a result of peripheral arteriolar dilatation.

Cardioprotective activity in primary prevention has never been demonstrated for the calcium antagonists, and apart from a subgroup in a small study with diltiazem, neither in secondary prevention following myocardial infarction. In the treatment of stable angina *isradipine* appears to be an effective compound, comparable with the other dihydropyridines. It can be used as monotherapy or together with a beta-blocker. The pattern of side-effects associated with the use of isradipine is probably very similar to that of the classical dihydropyridines.

With respect to a possible use of isradipine in *congestive heart failure* only few data are available, but clinical studies are in process. Short and medium term studies with other dihydropyridines in congestive heart failure reveal the modest and transient improvements to be expected from a drug-induced reduction of cardiac afterload. Studies with isradipine are ongoing, and its vascular specificity and weak or absent negative inotropic potency may be of some theoretical advantage. However, like for the other dihydropyridines, no effect other than arterial vascular dilatation is to be expected and the absence of dilatation in the venous bed is probably a disadvantage in the management of congestive heart failure. Since congestive heart failure is a complex syndrome with several possible backgrounds and mechanisms, a general recommendation or dismissal of a particular group of drugs is difficult and it can be imagined, that in certain subgroups of the disease a calcium antagonist, or more in particular isradipine, might be of some use.

Coronary dilatation in the large number of patients who simultaneously suffer from ischaemic heart disease and cardiac failure is a potential but clinically unproven advantage of the dihydropyridines.

5. So far, *Raynaud's phenomenon* has not been demonstrated to be influenced convincingly by dihydropyridine calcium antagonists, although some transient improvements have been reported incidentally. Since many of the trials reported in the literature are inconclusive because of methodological deficiencies, Thien performed a carefully designed study on the influence of nifedipine and nicardipine in both primary and secondary forms of the disease. However, only marginal and transient subjective improvements were established for both calcium antagonists and objective effects did not occur. The poor patient compliance may have been an additional problem in this and other studies. The therapeutic results so far obtained with calcium antagonists in patients with Raynaud's phenomenon are not particularly encouraging.

6. Calcium antagonists as potential *anti-atherogenic* drugs were reviewed by Weinstein. The data so far available have been obtained in various animal models of atherosclerosis. The anti-atherogenic potency of these drugs is not explained by any influence on the plasma lipoproteins, which remain virtually unchanged during treatment with calcium antagonists. An inhibitory influence of the calcium antagonists on collagen matrix synthesis and on cholesterol esterification may be the most likely mechanisms involved in the anti-atherogenic effects of the calcium antagonists.

It remains unknown whether the calcium entry blockade at the level of specific channels indeed underlies their anti-atherogenic activity. As a result of their lipophilicity the calcium antagonists so far studied are known to accumulate to an important degree in the cell membrane and may thus induce relevant biophysical changes in membrane structure and function, which may well be the basis of their anti-atherogenic effect.

In the animal experiments *isradipine* proved significantly more potent than nifedipine and other classical dihydropyridines, possibly as a result of isradipine's high lipophilicity.

The high anti-atherogenic potency of isradipine, established in animal experiments, has given rise to the initiation of the clinical trial MIDAS (*M*ulticenter *I*sradipine *D*iuretic *A*therosclerosis *S*tudy). In this ongoing trial hypertensive patients with small early lesions of a carotid artery (as established by ultrasound Doppler analysis) are treated with either isradipine (10 mg dd.) or with hydrochlorothiazide. The patients treated with the diuretic will only receive additional enalapril if necessary for blood pressure control. The results will be known in the early nineties.

7. Data obtained during the last few years have confirmed the position of isradipine as a dihydropyridine with a high degree of vascular specificity and very weak negative inotropic activity. Future studies should include, apart from the treatment of hypertension and angina, also its influence on the microcirculation, on atherogenesis and possibly also on cerebral ischaemia.